Open Systems Dependability

Dependability Engineering for Ever-Changing Systems

Second Edition

Open Systems Dependability

Dependability Engineering
for Ever-Changing Systems

Second Edition

Editor

Mario Tokoro

Co-founder, Executive Advisor
Sony Computer Science Laboratories, Inc., Tokyo, Japan

Formerly Sr. Vice President and CTO
Sony Corporation, Tokyo, Japan

CRC Press
Taylor & Francis Group
Boca Raton   London   New York

CRC Press is an imprint of the
Taylor & Francis Group, an **informa** business

A SCIENCE PUBLISHERS BOOK

CRC Press
Taylor & Francis Group
6000 Broken Sound Parkway NW, Suite 300
Boca Raton, FL 33487-2742

International Standard Book Number-13: 978-1-4987-3628-2 (Hardback)

Visit the Taylor & Francis Web site at
http://www.taylorandfrancis.com

and the CRC Press Web site at
http://www.crcpress.com

PREFACE

Technological advancements in the 20th century—particularly in the fields of electronics, computers, the internet, and mobile communications—have drastically expanded our capabilities from the twin perspectives of time and space. Industries have developed and the world economy has grown. Our standard of living has improved significantly in terms of quantity and quality thanks to the support of services made possible by these advancements. In the recent past, the functionality of services has also expanded, especially where realized in conjunction with sensors and actuators, and the underlying systems, which are fused with the real world, form an integrated infrastructure that supports our everyday life. Ultimately, our daily lives have taken on a new dimension never experienced before.

On any morning you could wake up, switch on your TV set, check your e-mail, eat breakfast, and leave your house for work. TV shows are transmitted under the control of computers. The electricity that runs your TV set and appliances is also distributed under computer control. Traffic lights are synchronized by computers in order to ensure that traffic flows smoothly. Electronic ticketing cards for public transit rail are controlled by computers that calculate your fare by detecting where you enter and leave the system. Train operation is, of course, also controlled by computers.

These systems operate for long periods of time without any interruption to services. During this lifetime, however, their purpose and the environment in which they operate may change. For example, the Suica® system of electronic fare cards operated by the East Japan Railway Company was initially used for commuting pass, but its scope has now been extended to include regular rail tickets, taxi fares, and kiosk purchases; in addition, other railway companies are also now using similar electronic fair cards interoperable with Suica. Systems such as this operate impassively, while unbeknownst to most users, they undergo revision and modification to accommodate changes in their purpose and environment. There are almost certainly numerous examples of this type of system very close to you as you read this book.

However, once a problem occurs in a system and services are interrupted, we are greatly inconvenienced, and only then become aware that how our daily lives depend upon them. We therefore need to think about how we can make these systems dependable in order to prevent similar inconvenience in the future.

Fields such as reliability engineering and dependability engineering have studied problems in systems, but their methods target systems with specifications that do not change over time. This assumption did work for the systems of the past, i.e., those that were independent, not connected to one another, and rarely saw changes in purpose or environment. Today's systems are, however, interconnected and always changing. Therefore, there is a pressing need for a concept and methods that will allow dependability to be attained in ever-changing systems.

As the second edition of *Open Systems Dependability—Dependability Engineering for Ever-Changing Systems*, this book answers that need. It describes systematically the concept and methods for reducing problem incidences, avoiding serious accidents, supporting accountability, and ensuring continuous operation in a large and complex system that runs over a long period of time. A system having a boundary, functions, and a structure that are all subject to change is called an *open system*. We refer to the concept as *Open Systems Dependability* and to the technical system comprising the concept and methods as *Dependability Engineering for Open Systems*, or DEOS for short.

We hope that this book will help to ensure greater dependability in future information systems, and in doing so, will make our modern society safer, more secure, and more convenient.

January, 2015   **Mario Tokoro**

ACKNOWLEDGEMENTS

This book is based on the DEOS Project supported by the Core Research for Evolutional Science and Technology (CREST) program of the Japan Science and Technology Agency (JST) under the Ministry of Education, Culture, Sports, Science and Technology in Japan (MEXT). I am indebted to JST and MEXT for their strong support for the DEOS Project. I would like to extend my gratitude to Dr. Koichi Kitazawa (then representative director of JST), Dr. Toshiaki Ikoma (then head of the Center for Research and Development Strategy), and Mr. Shigeru Ishimasa (then Director of Research and Development Strategy) for their understanding of the importance of this project and for various arrangements that allowed it to proceed. I would also like to thank Mr. Shigeki Sakai, Mr. Takashi Yakushiji, Mr. Masanori Emori, and Mr. Katsumi Takeda of JST for their strong administrative support.

I am very grateful to Co-Research Supervisor Professor Yoichi Muraoka and all of the Area Advisors from the DEOS project—Dr. Kazuo Iwano, Prof. Toru Kikuno, Dr. Koichi Matsuda, Prof. Koichiro Ochimizu, Dr. Yoshiki Seo, Prof. Hideki Tanaka, and Prof. Hiroto Yasuura—without whose advice this project could not have been successful. I also thank the Research Promotion Board members—Mr. Nobuhiro Asai, Mr. Shingo Kamiya, Mr. Tadashi Morita, Dr. Masamichi Nakagawa, Mr. Takeshi Ohno, Mr. Ichiro Yamaura, and Dr. Kazutoshi Yokoyama—for their close communication with research teams in order to ensure that the project's findings and deliverables could be put to practical use. I am indebted to the Area Management Advisors—Mr. Kazuo Kajimoto, Prof. Yuzuru Tanaka, Mr. Tetsuya Toi, Prof. Seishiro Tsuruho, and Dr. Daiji Nagaoka—for their advice from the industrial and application perspectives. And I would also like to thank Mr. Makoto Yashiro, Director of the DEOS R&D Center, as well as Mr. Shigeru Matsubara, Dr. Hiroki Takamura, and all of the other members of the center for their daily support and for integrating the research results into actual processes.

I am grateful to our external reviewers—the late Dr. Jean-Claude Laperie, Prof. Robin Bloomfield, Prof. Jean-Charles Fabre, Mr. Masayuki Hattori, Dr. Karama Kanoun, and Prof. Miroslav Malek—for their perceptive comments and advice throughout the course of this project. I would like to give my personal thanks to Mr. Junkyo (Jack) Fujieda for his valuable advice on standardization. I also thank Mr. Matthew Heaton and Mr. Paul O'Hare of Translation Business Systems Japan (TBSJ) for their help in accurately translating this book from Japanese into clear and comprehensible English.

Last but not least, I thank all of the contributors to this project—Prof. Yutaka Ishikawa, Dr. Satoshi Kagami, Prof. Yoshiki Kinoshita, Prof. Kenji Kono, Prof. Kimio Kuramitsu, Prof. Toshiaki Maeda, Prof. Tatsuo Nakajima, Prof. Mitsuhisa Sato, and Prof. Hideki Tokuda—for their team leadership, support for the DEOS core team, and hard work in both research and team management. I am also grateful to all members of each of these teams. And as we publish this second edition, it is with great pleasure that I extend my most sincere thanks to my fellow authors.

CONTENTS

AUTHORS PER CHAPTER/SECTION

Chapter	Section	Authors
1		Mario Tokoro
2	2.1	Mario Tokoro, Hiroki Takamura
	2.2	Mario Tokoro, Shigeru Matsubara, Hiroki Takamura
	2.3	Mario Tokoro
	2.4	Mario Tokoro
3	3.1	Mario Tokoro
	3.2	Mario Tokoro
	3.3	Mario Tokoro
	3.4	Mario Tokoro, Shigeru Matsubara
4	4.1	Shuichiro Yamamoto
	4.2	Yutaka Matsuno
	4.3	Yutaka Matsuno
	4.4	Yutaka Matsuno
	4.5	Shuichiro Yamamoto
	4.6	Shuichiro Yamamoto
5	5.1	Atsushi Ito
	5.2	Hideyuki Tanaka
	5.3	Yutaka Matsuno
6	All	Makoto Takeyama
7	7.1	Yasuhiko Yokote, Kiyoshi Ono, Tomohiro Miyahira
	7.2	Kiyoshi Ono, Tomohiro Miyahira
	7.3	Satoshi Kagami
	7.4	Kenji Kono, Hiroshi Yamada
8	All	Kimio Kuramitsu
9	All	Yasuhiko Yokote, Tatsumi Nagayama, Sachiko Yanagisawa
10	All	Yoshiki Kinoshita, Makoto Takeyama
11	All	Mario Tokoro
Appendices	A.1	Makoto Yashiro, Shigeru Matsubara
	A.2	Makoto Yashiro
	A.3	Hiroki Takamura
	A.4 to A.6	Shigeru Matsubara

INTRODUCTION

In today's modern world, our daily lives would have been almost impossible without the help of information systems. We receive tremendous benefit from this type of systems in the form of mobile phones and Internet-based services, but there are countless other examples, such as weather forecasting, services provided by the fire departments, the police, and other similar organizations. Infrastructure for telecommunications, broadcasting, and the power grid would not be possible without information systems. In addition, reservation management systems, traffic monitoring, and automatic turnstile and check-in terminals for trains and airplanes rely on this technology, as do logistics systems, commercial systems for banking and financing, and corporate management systems. These information systems are always online, and they provide services in real time. In recent years, they have become increasingly interconnected—either directly or indirectly—to form even larger systems that serve as the bedrock of our modern lifestyles. As we grow more and more dependent upon these systems, it is evermore important that they be as reliable, robust, and safe as possible. In this book, the term *dependability* is used to refer to these and other information-system attributes that allow users to avail of their services continuously and securely.

The traditional approach adopted in developing information systems has involved detailed preliminary planning, meticulous description of the required system specifications, and an exhaustive course of design, implementation, testing, and verification before they could finally deliver products and services to end users. This method has proved very effective with products and services for which system requirements can be clearly defined at the start of development and system specifications covering the entire service life can be accurately estimated in advance. However, it seems this is not enough nor appropriate for today's systems.

Today's systems are increasingly required to provide a much broader range of services, to be much larger in scale in order to satisfy these needs, and to also operate continuously over extended periods of time in real-world situations. As a result, service objectives and the requirements of users often change over the course of an information system's service life, and service providers must accommodate these changes without disturbing the operation of the system. In addition, the service providers are required to update these systems constantly in response to advances in technology, changes in the legal system, and revisions in international standards. Clearly, system specifications that satisfy all possible requirements at any time in the future cannot be determined at the initial stages of development. We are thus faced with a situation wherein the system development process itself must be overhauled in order that information systems can accommodate change.

Yet information systems provide critical support for our daily lives, and therefore, it is equally as important that they remain dependable. This is particularly true in the case of interconnected systems that serve as the infrastructure for modern lifestyles, as a fault in any one could easily propagate to some or all of its connected systems. Meanwhile, developers often reuse existing code and purchase over-the-counter software products in order to reduce development time and cost, and as a result, it is inevitable that software components having unknown specifications or developed in line with different standards will be combined with one another. Cases where administrative responsibility is spread across multiple domains—such as with services provided via networks and cloud-based systems—would also seem to be on the increase. In all of these cases, information system developers must satisfy the dependability requirements of individual customers and of society at large. And in order that their systems may be used with confidence, they are also required to address appropriately the risk of system attack by malware, disclosure of information due to unauthorized access, and other similar threats [1].

Against this backdrop, research and development are aimed at assuring dependability is underway at a frenetic pace in countries all over the world, and many standards and guidelines are being published as a result. Some information systems are being built in lines with these standards and guidelines, but they are yet to be widely adopted. Furthermore, the standards and guidelines have not themselves been developed to such a level where they can ensure dependability of modern information systems with the characteristics and attributes described above. Unfortunately, critical systems continue to fail all over the world while we wait (see Appendix A.3).

Analysis of such failure reveals a number of different causes. In some cases, it was impossible to determine accurately the functions and conditions of use of all of the component elements when creating the architecture; in others, the load on the system or the number of people using it had exceeded the original estimates. With some systems, failure occurred because they were no longer suitable for use due to their age or an inability to identify inconsistencies introduced when they were updated to accommodate changes in requirements. Failure of an information system significantly inconveniences its users, but from the perspective of service providers, it can also result in lost business, compensation payouts, damage to their brand, or even the discontinuation of business. For this reason, dependability and further enhancements are one of the biggest challenges faced by these companies.

Meanwhile, there also remain the fundamental concerns as to whether large, complex systems that will be put to long-term use in real-world situations and must, therefore, continually accommodate change, can actually be built in a perfect, flawless manner. Recently developed information systems, as well as those of the future, cannot be seen as having rigidly defined functionality, structures, or boundaries. From the standpoints of both development and application, it is more appropriate to assume that they will constantly evolve over time [2]. We have thus come to the conclusion that making information systems capable of (a) accommodating changes in requirements, (b) eliminating the causes of potential problems before they can lead to failure, (c) minimizing the scale of damage if failure does occur, and (d) assisting service providers and other stakeholders in achieving accountability will be highly effective in assuring dependability. This represents an iterative approach that aims at assuring the dependability of constantly evolving systems based on accountability. We refer to our asymptotic approach as *Open Systems Dependability* (OSD) in this book in order to distinguish it from others who seek to build the perfect system at the development stage [3].

Based on our new approach, this book will describe the basic concepts, techniques, and fundamental technologies that are critical for assuring dependability in massive, complicated systems required to operate for long periods and continuously accommodate change. We also cover the software developed for this purpose, case studies, and the current state of progress in terms of standardization, among other issues.

REFERENCES

[1] Yasuura, H. 2007. Dependable Computing for Social Systems, Journal of IEICE, Vol. 90, No. 5, pp. 399 to 405, May 2007.
[2] Tokoro, M. (ed.). 2010. Open Systems Science—From Understanding Principles to Solving Problems, IOS Press.
[3] Tokoro, M. (ed.). 2012. Open Systems Dependability—Dependability Engineering for Ever-Changing Systems, CRC Press.

2

OPEN SYSTEMS DEPENDABILITY

2.1 EVOLUTION OF APPROACH

Let us start by looking back over the history of dependability. During the Cold War in the 1960s, as the Apollo program aimed at putting man on the moon, the fault tolerant computer was proposed as a means of supporting real-time computing and mission-critical applications. This marked the start of active discussion in the field of dependability [1, 2]. The subsequent increase in scale of hardware and software and the increased popularity of online services led to the development of RAS—Reliability, Availability, and Serviceability. Focusing on system-error detection and recovery, this concept integrated resistance to failure (*reliability*), assurance of a high operation ratio (*availability*), and the ability to rapidly recover in the case of failure (*serviceability* or *maintainability*) [3, 4]. As computers started to be used more and more for business in the latter half of the seventies, two additions were made to RAS—namely, assurance of consistency in data (*integrity*) and the prevention of unauthorized access to confidential content (*security*). This extension of the original concept, RASIS, has provided a standard for the evaluation of information systems. With the turn of the century, the concept of autonomic computing was proposed as a means of achieving the highest possible level of self-sustained dependability in complex systems connected by networks, and this approach took its inspiration from the human involuntary nervous system [5, 6, 7, 8] (Fig. 2-1).

There has also been considerable progress in terms of the methods used to develop software, which is a critical aspect of information system dependability. First to be introduced were programming methods such as structured programming (Dijkstra [9]) and object-oriented programming (Simula [10], Smalltalk [11]). The field subsequently grew to include management methods for software development projects, before the focus shifted to development processes for software (CMM [12, 13], CMMI [14]). Projects dealing with development methods for complicated, large-scale

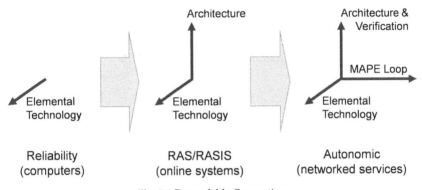

Fig. 2-1 Dependable Computing.

systems (system of systems, ultra-large-scale systems [15]) also got underway. Development approaches have also changed: whereas the waterfall model was the most popular for many years from the seventies onwards, 2000 saw the arrival of new methods such as agile software development [16] and DevOps [17] (Fig. 2-2). In addition, COBIT [18], ITIL [19], and other best-practice frameworks were also introduced for IT governance and service management in organizations such as companies and administrative bodies.

International standards reflect the way in which approaches taken to ensure reliability and security have changed (Fig. 2-3). The IEC 60300 international standards are well known in the field of dependability management. Because Technical Committee 56, which is responsible for dependability-related standards at the IEC, originally dealt with the reliability of electronic components, software dependability is not adequately covered

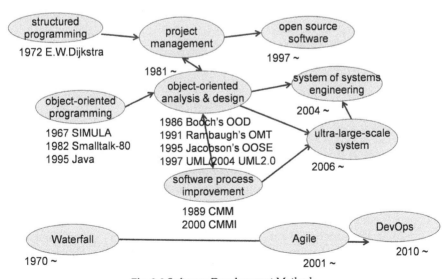

Fig. 2-2 Software Development Methods.

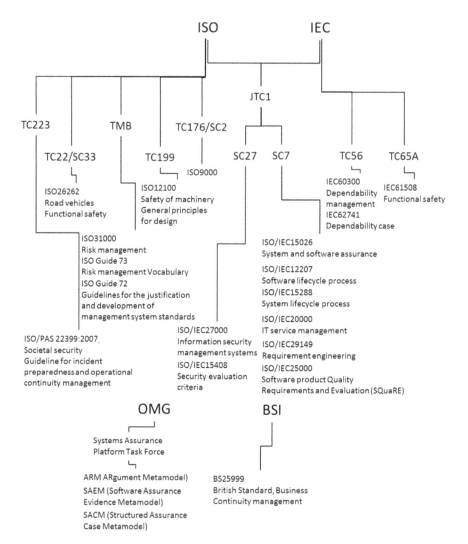

Fig. 2-3 Related International Standards.

by IEC 60300-1 (2003 edition)—a key standard within the IEC 60300 series. Revision work is therefore currently underway and it is planned to expand greatly the scope of dependability management to include products, systems, services, and processes. International safety standards ISO 13849-1 (EN954-1) and IEC 60204-1 deal with simple components and devices, and as such, do not cover systems that include software.

In 2000, the functional safety standard IEC 61508 was established out of necessity in order to support this type of system (Fig. 2-4). The standard classifies device failure as either random hardware failure or systematic

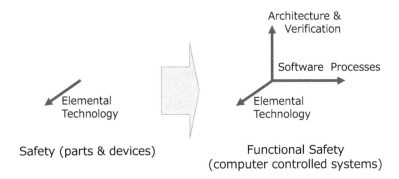

Fig. 2-4 Functional Safety.

failure. The probability of random hardware failure is determined based on the deterioration of components, while in the case of systematic failures, problems arising from the system's design, development, manufacture, maintenance, and operation are to be kept within permitted ranges through software verification using procedures and documentation based on the safety lifecycle, the V-model, and other techniques. In addition, systems designed, developed, and manufactured in this way are categorized as having either a low-demand operation mode or a high-demand or continuous operation mode. Target levels of risk reduction are defined for each mode, and these are managed in the form of safety integrity levels (SILs). Four requirement levels have been established—SIL 1 to SIL 4—with the latter having the strictest requirements in terms of safety function. A range of other standards were developed on the basis of IEC 61508, such as IEC 62061 for machinery, IEC 61511 for processes, IEC 61513 for nuclear power, and IEC 62278 for railways; in addition, ISO 26262 was published for the automotive industry in 2012. Compliance with ISO 26262 requires the development of a safety case, which is a structured argument based in turn on the assurance case, and the ISO/IEC 15026 international standards also highlight the importance of assurance cases from the viewpoint of system assurance.

Meanwhile, efforts to encapsulate the various perceptions of dependability in a single definition continue. IFIP Working Group 10.4, which is concerned with dependable computing and fault tolerance, and the IEEE Technical Committee on Fault Tolerant Computing initiated a joint project in 1980 to research the fundamental concepts and terminologies of dependability. The details and findings of the investigation that followed were compiled in a technical paper published in 2004 [20, 21]. In this technical paper, dependability and security are defined as concepts that are different but interrelated: dependability consists of availability, reliability, safety, integrity, and maintainability; while security consists of availability, confidentiality, and integrity (Fig. 2-5).

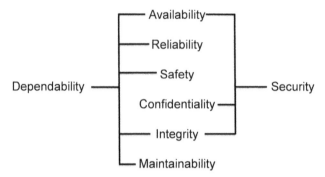

Fig. 2-5 Dependability and Security defined by IFIP WG 10.4.

In this book, we define dependability as *the ability of a system to continuously provide services expected by users*. The above-mentioned properties are all constituents of dependability. Robustness and resiliece can also be considered constituents. As dependability requirements may differ by application domain, or even by system, the importance of each property may differ. We treat dependability as the integrated concept consisting of related properties.

Yet, despite ongoing research in the field of dependability and technological advancements based on this work, large-scale information systems continue to fail. Some notable examples of these failures are presented in Appendix A.3. Analysis has shown the most noteworthy causes to be continued development of the system architecture even though the component elements were not fully understood; system parameters such as the number of users, the transaction volume, the amount of data, and processing ranges exceeding the initial design levels; and the introduction of inconsistencies into system operation due to addition or modification of functionality in response to changing user requirements. In certain other cases, carelessness on the part of programmers or operators caused a cascade of failures, ultimately taking the entire system offline.

Our approach to the concept of dependability has changed significantly over the years in response to the needs of the times. However, none of the approaches adopted till date can adequately deal with the systems we have identified—that is, large-scale systems that can be efficiently adapted to ever-changing objectives and environments while continuing to provide the services required by their users. We need to see these systems as integrated wholes with no distinct boundaries between development and operation. We cannot assume that information systems will—from the perspective of development and operation—always retain the same clearly defined system boundaries, functions, and structure throughout their lifetimes. Instead, we must assume that these aspects of the system will evolve with the passage of time. In the following sections, we will start by clarifying the characteristics of today's information systems and the major causes of their failure. We will then present a new concept of dependability for these modern systems.

2.2 CHARACTERISTICS OF TODAY'S SYSTEMS AND CAUSES OF THEIR FAILURE

In order to shorten the development period of large-scale information systems and to reduce the cost involved, the reuse of existing code (legacy code) or software products from other companies as black-box programs is on the increase (Fig. 2-6).

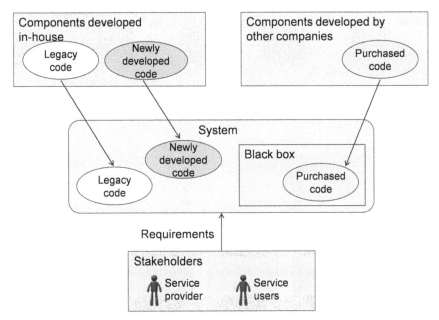

Fig. 2-6 Various Components at Development.

In addition, a system's specifications often need to be changed while it is in operation so that the services provided can be improved or otherwise modified. Development for specification changes can be undertaken by various methods. In recent years, demand has been growing for such changes to be made without interrupting the services that the system provides. In response, it is not unusual for maintenance personnel to implement modifications manually or for updates to be downloaded via a network. Especially when frequent changes of this nature make it extremely difficult for designers, developers, and operators to understand every aspect of a system at all stages of its lifecycle, from development through to retirement (Fig. 2-7).

Many of the information systems in operation today are connected to other systems via networks in order to provide extended services. In this type of situation, while users directly utilize services from one service domain, they indirectly use those from others too. It is often the case that these service domains are owned and operated by different companies. As such, there

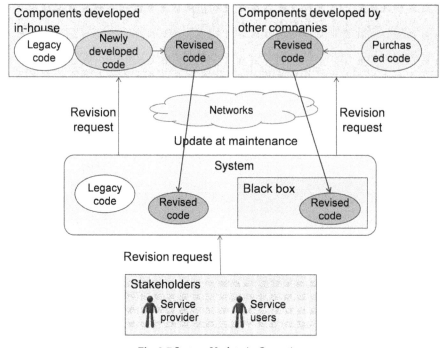

Fig. 2-7 System Update in Operation.

is always a possibility that service items, content, transactions, interface specifications, and the like could be changed without sufficient notice being given, that unfamiliar services could be added, or that certain services could even become unavailable. Given that the same also applies to networks, users and system developers are finding it increasingly difficult to identify the boundaries between systems and also between service domains. To make matters worse, hackers can attack systems at any time with malicious intent. For all of these reasons, predictability has become much more difficult with the advent of networking (Fig. 2-8). When categorized from the viewpoints of system development and operation, we can see that system failure can be caused by either (1) the system itself being incomplete or (2) the environment in which the system exists being uncertain.

(1) System Incompleteness

For several reasons, it is extremely difficult to build a perfect system in an environment such as the one described above. For example, the actual requirements of the system may be ambiguous to some degree, and when combined with natural language, this can result in a difference of understanding between the ordering party and the developer. Furthermore,

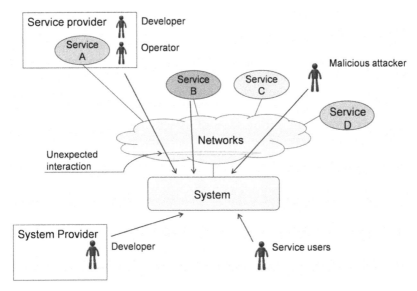

Fig. 2-8 System including External Services through Networks.

when the requirements are documented in the form of specifications, it can be difficult to ensure that absolutely nothing is omitted and that the specifications themselves are reflected perfectly in the developed system. Particularly in the case of systems put to use over a prolonged period of time, this type of incompleteness can never be eradicated, despite repeated modifications to the system in order to accommodate changes in environment as described below.

Some more specific examples of factors causing failure are:

- The system being configured from multiple different pieces of software, making inconsistencies in the specifications more likely and also rendering exhaustive description of the specifications and testing impossible should the system grow in size or complexity;

- Errors or omissions in the deliverables of development phases such as requirement elicitation, drafting of specifications, design, implementation, and testing due to the corresponding team members not sharing a common understanding, poor documentation, and so forth;

- A mismatch between specifications and the actual operation of blackbox programs or legacy code;

- Errors related to modification or repair during administration, operation, or maintenance; and

- Expiration of licenses.

(2) Uncertainty in the System's Environment

The environments in which information systems operate are in a constant state of flux, yet when designing and developing this type of system, understanding the environment is a critical precondition. Designers and developers must therefore make assumptions about the operating environment over the planned lifecycle of the system, but at this early stage, it is simply not possible to accurately foresee the future state of the environment. In other words, changes to the system must be implemented after it goes into operation in order to compensate for this uncertainty surrounding its environment. And when making these changes any time during the lifecycle of the system, incompleteness can be introduced, potentially leading to system failure. Some possible causes of such failure are:

- New system requirements when business objectives are altered;
- Changes in user requirements or what is expected of the system;
- Changes in use cases due to increasing numbers of shipments or users and changes in capacity economics;
- Technological innovation;
- Updates to new or revised standards and regulations;
- Changes in operator abilities or proficiency;
- Unforeseeable connections made possible by the network environment; and
- Deliberate attacks from outside the system.

To summarize, today's large-scale information systems must constantly deal with incompleteness in their own implementation and uncertainty in their environment while they continue to operate (Fig. 2-9).

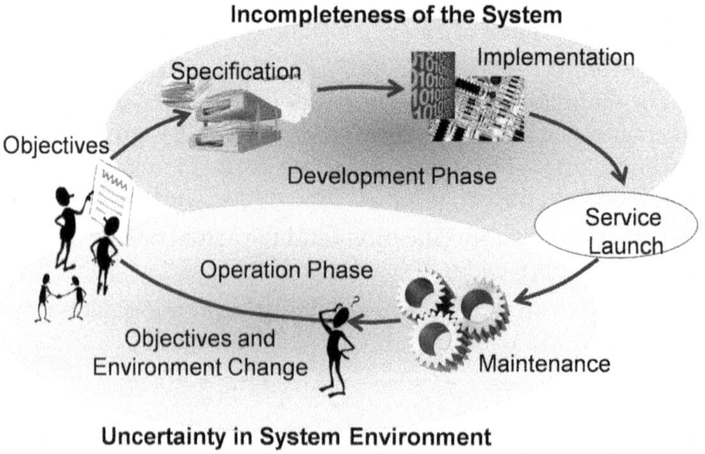

Fig. 2-9 Incompleteness and Uncertainty.

In light of this situation, many attempts have been made to define the concept of dependability. Two notable examples are: 'The ability—despite complete non-occurrence of failure and malfunction being preferable—to immediately ascertain the situation in the event of an abnormality, to predict how the situation will develop from that point onward, and at a reasonable cost, to prevent social panic and catastrophic breakdown' [22] and 'The ability to continue to provide system services at a level acceptable to users if accidents occur' [23].

While it may be impossible to completely avoid all potential forms of failure, we believe it is possible to develop methods and technologies that reduce occurrences of critical failure to the minimum and that allow business to continue in the event of such a failure by minimizing the scale of damage suffered, preventing the same failure from occurring again in the future, and achieving accountability. This is the ultimate aim of our work, and in accordance with a new definition of dependability as presented in the following section, we strive to develop the methods and technologies needed to achieve it in practice.

2.3 CONCEPT AND DEFINITION OF OPEN SYSTEMS DEPENDABILITY

Our research is focused on the type of large, complex system that must operate for extended periods of time in real-world settings containing people and must, therefore, be able to accommodate change on an ongoing basis. Thus, for the reasons described above, it is not possible to completely eliminate incompleteness in specifications and implementation or uncertainty caused by changing user requirements and operating environments. As opposed to a closed system, a system of this nature can be labeled an *open system*. To better explain the differences between these two types of system (Fig. 2-10), the characteristics of each are presented below.

The general features of closed systems are:

- The boundaries of the system can be clearly defined;
- The functionality of the system does not change; and
- The structure and component elements of the system are fixed, as are the ways in which the component elements interact with one another.

The following can be said of these features:

- An observer can view the system from the outside; and
- Reductionism is applicable.

Closed System   **Open System**

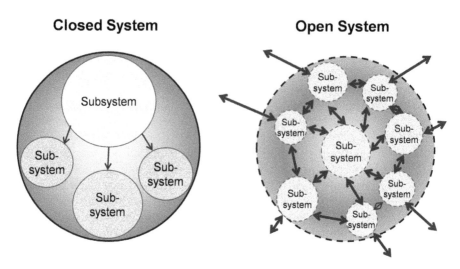

Fig. 2-10 Closed System vs. Open System.

In contrast, the general features of open systems are:

- The boundaries of the system cannot be defined;
- The functionality of the system changes over time; and
- The structure and component elements change over time, as do the ways in which the component elements interact with one another.

We can derive the following based on these features:

- The observer is also included within the system and can only view it from the inside; and
- Reductionism is not applicable.

All real-world objects undergo change while simultaneously dispersed and exerting an influence on one another. Accordingly, all systems existing in the real world constitute an open system in which they are interconnected. That said, if we were to extract one portion of this overall system and temporarily remove all interaction with the others, it would become much easier to understand its fundamental concept and principles. This idea that a portion of the real world environment can be interpreted by isolating it is known as the *closed system hypothesis*. In more specific terms, if we extract a portion for which the closed system hypothesis holds, or alternatively, if extraction is possible in such a way that the hypothesis holds, we can easily determine the fundamental concept and principles of that portion of the overall system. Having contributed greatly to science after being discovered by Descartes in the 17th century, the methodology of reductionism is used to understand problems in a complex system by successive reduction to the smallest possible units followed by recombination [24]. Particularly, in the field of physics, where one specific portion or property can easily be extracted and interpreted, this approach in combination with the advancement of mathematics has paved the way for remarkable progress. Meanwhile, in medicine, biology, agriculture, and other natural sciences which study items living in the real world as well as in social sciences, including economics, it is not as easy to extract an individual portion from the relevant complex system (or there is opposition to such action). For example, Bertalanffy defined an open system as a 'system in exchange of matter with its environment, presenting import and export, building-up and breaking-down of its material components' [25]. It is debatable, therefore, whether reductionism alone can serve as a basis for understanding and problem-solving.

In the field of engineering, where deliverables are implemented in a real-world environment after reduction (analysis) and recombination (synthesis), success is measured not only based on the desired effect having been accomplished, but also on the absence of any negative effect on the target system and its environment beyond permitted limits. Therefore, consideration must always be given to those interactions with other portions of the system that are temporarily eliminated as a result of extraction.

In scientific terms, the study of information systems is a field of engineering. Historically, development in this area began with simple offline systems for which the extent of the influence on and by other systems could be regulated, and mathematical or constructive methods work well with this type of system. From this start point, the scale of software has increased greatly and the number of online applications has grown; in addition, software-process methods came to see people as one constituent element of the overall system. Meanwhile, although not explicitly stated, the adoption of an open-system approach would now seem to be indispensable in the development of systems-of-systems and ultra-large systems. In the past few years, international standards dealing with dependability have finally started to reflect this situation.

Let us return to the type of system that is the subject of our research—large, complex systems that must operate for extended periods of time in real-world settings containing people and must, therefore, be able to continuously accommodate changes in the objectives of the services they provide, in the needs of their users, in technology, and in standards and related legislation. Because of these requirements, the functionality and structure of these systems often evolve more than originally foreseen, which also causes their boundaries to change. Furthermore, in cases where services provided by external systems are employed or some of its functionality is hosted externally in the cloud, this type of system may ultimately operate across several administrative zones, making it literally impossible to clearly identify its boundaries. In other words, the systems that we study plainly exceed the general nature of the closed system, and for this reason, reductionism is often not applicable. Meanwhile, we consider all of the people associated with the development and operation of such a system—be they the owners, designers, developers, operators, users, or others—to be internal observers, because of the effect they have on it. Accordingly, it is our position that these systems should be accepted as being open systems without hesitation so that the concept of dependability for open systems may be established.

One could interpret this type of system as being closed at certain times, allowing it to be treated as a definable system undergoing no change over time. Overall dependability could then be studied through concatenation of these snapshots. This specific approach has, in fact, been principally adopted in past development of dependable systems. For this purpose, the boundaries, functionality, and specifications of the system must be clearly defined for each point in time, and the processes of design, verification, and testing must then be carried out in a repetitive cycle based on these definitions. However, a large system is normally composed of many subsystems, more than one of which are modified and adapted simultaneously and in multiple ways without interrupting the services provided, and this makes it extremely difficult to distinguish periods where a system is changing from those in which it is not. Particularly when the system under consideration has a distributed nature, it is impossible to obtain a unique view thereof [26, 27]. In such a case, the approach of distinguishing between fixed periods and change periods simply cannot be applied in practice.

This being the case, we should instead focus on ever-changing systems and establish a concept of dependability with the main aim of ensuring that, in this environment of constant flux, services and business can continue without interruption in the event of system failure and accountability can be achieved. In other words, we must focus our research on open systems and adopt an approach that enhances dependability as much as possible in a recursive process over the lifecycle of the system. On this basis, we have established the following definition for the *Open Systems Dependability (OSD)* required of information systems, both now and in the future.

> *Open Systems Dependability is the ability of a system operated for an extended period of time in a real-world environment to accommodate change in its objectives and environment, to ensure that accountability in regard to the system is continually achieved, and to provide the expected services to users without interruption.*

The prime assumption for OSD and this definition can be structured as follows:

1. *Assumption*: Potential causes of failure cannot be completely eliminated from a system operated for an extended period of time in a real-world environment.
2. *Definition*: OSD is achieved when a system is capable of:
   i) Continually accommodating change in the objectives and environment thereof;
   ii) Ensuring that accountability is continually achieved; and
   iii) Providing the expected services to users without interruption.

The above assumption reflects a fact of life that cannot be ignored. The ability of the system to continually accommodate changes in its objectives and environment as stated in the definition primarily benefits service and product providers, and some or all of the other stakeholders. The ability of the system to ensure achievement of accountability also primarily benefits service and product providers, but also other stakeholders, including the users of the system, and ultimately society at large. It is the users who chiefly enjoy the ability of the system to provide continuous services as mentioned in the definition, but all other stakeholders including service and product providers also profit as a result.

Even if a system has the capabilities described above, there is no guarantee that failure will never occur. As stated in the assumption, this is a fundamental characteristic of the open system. Accordingly, OSD is enhanced through an iterative process of providing the necessary functionality for eliminating the causes of failure to the greatest extent possible, minimizing the scale of damage in the event of failure, maintaining accountability, and preventing any failure from reoccurring. The next section will discuss in detail the key principles for achieving this.

2.4 TOWARD THE REALIZATION OF OPEN SYSTEMS DEPENDABILITY

In order that an information system may satisfy the requirements of OSD in practice, it will likely need a set of specific functions as described below:

First of all, in order that a system can accommodate changes in its objectives and environment, the system must have functionality for modifying itself in response to the corresponding change requirements. Before this, the stakeholders must agree to the change requirements and the methods for realizing the necessary changes. Design and development work must then be carried out in order to reflect this agreement. We refer to this function as *change accommodation*.

Next, let us look at the ability of the system to provide the expected services to users in as continual a manner as possible. In light of our assumption that potential causes of failure can never be completely eliminated, the system needs an additional three functions—*failure prevention, responsive action*, and *recurrence prevention*—in order to satisfy this requirement. Failure prevention refers to functionality for removing potential causes of failure to the greatest extent possible before they can cause damage. Responsive action, meanwhile, is functionality that makes it possible to respond quickly and appropriately to any failure and to minimize the damaging effect thereof. And as its name suggests, the functionality referred to as recurrence prevention facilitates analysis of the causes of a failure that have unfortunately developed *(cause analysis)* and modification of the system so that it will never again fail due to the same causes.

Accountability is mainly needed when a system failure has occurred. To ensure continuous achievement of accountability, the *cause analysis* functionality described above must be used. And in order to analyze accurately the causes of a failure, the following two functions—*agreement log retention* and *operation log retention*—are required. As we have seen, whenever a system must accommodate changes in its objectives or environment, the stakeholders must agree on the new requirements placed on the system and how the changes will be realized: the function referred to as agreement log retention makes it possible to notate these arguments in a structured manner and to also maintain logs describing the process of agreement. Operation log retention is functionality that monitors the operation status of the system and records what it finds.

In order that the functions identified above can be realized on a continuous basis, they must be implemented within the system under consideration by means of an iterative process.

Focusing on the development of technologies for enhancing system safety and security, previous research in the field of dependability has principally targeted accidental and incidental faults, viewing a system as a fixed one. In contrast, we developed our approach for systems affected by change over time, aiming to provide them with the ability to accommodate such change. In this regard, we centered our efforts on open system failure resulting from incompleteness and uncertainty. Furthermore, our objective has been to enhance system dependability by providing functions that facilitate continuous system operation as well as functionality to ensure accountability achievement, and also by implementing, within the system itself, an iterative process that combines these functions. This approach to dependability is a fresh departure from those of the past. In the next chapter, we will take a detailed look at the DEOS Technological System established in line with this key principle.

REFERENCES

[1]   Koob, G.M. and C.G. Lau. 1994. Foundations of Dependable Computing, Kluwer Academic Publishers.
[2]   Avizienis, A. 1967. Design of fault-tolerant computers. In Proc. 1967 Fall Joint Computer Conf., AFIPS Conf. Proc. Vol. 31, pp. 733–743.
[3]   Hsiao, M.Y., W.C. Carter, J.W. Thomas and W.R. Stringfellow. 1981. Reliability, Availability, and Serviceability of IBM Computer Systems: A Quarter Century of Progress, IBM J. Res. Develop., Vol. 25, No. 5, pp. 453–465.
[4]   Diab, H.B. and A.Y. Zomaya. 2005. Dependable Computing Systems, Wiley-Interscience.
[5]   Ganek, A.G. and T.A. Corbi. 2003. The dawning of the autonomic computing era, IBM Systems Journal, Vol. 42, No. 1, pp. 5–18.
[6]   An architectural blueprint for autonomic computing, 4th edition, IBM Autonomic Computing White Paper, June 2006.
[7]   http://www-03.ibm.com/autonomic/
[8]   Huebscher, M.C. and J.A. McCann. 2008. A survey of Autonomic Computing, ACM Computing Surveys, Vol. 40, No. 3, Article 7, August 2008, pp. 1–28.
[9]   Dahl, O.J., E.W. Dijkstra and C.A.R. Hoare. 1972. Structured Programming, Academic Press, London.
[10] Birtwistl, G.M. 1973. SIMULA Begin, Philadelphia, Auerbach.

[11] Smalltalk: http://www.smalltalk.org/main/
[12] Humphrey, W. 1988. Characterizing the software process: a maturity framework. IEEE Software 5 (2), March 1988, pp. 73–79. doi:10.1109/52.2014. http://www.sei.cmu.edu/reports/87tr011.pdf
[13] Humphrey, W. 1989. Managing the Software Process, Addison Wesley.
[14] http://www.sei.cmu.edu/cmmi/
[15] Ultra-Large-Scale Systems: The Software Challenge of the Future, http://www.sei.cmu.edu/library/assets/ULS_Book20062.pdf.
[16] Beck, K. 1999. Extreme Programming Explained: Embrace Change—First Edition, Addison-Wesley, Boston.
[17] Loukides, M. 2012. What is DevOps? Infrastructure as Code, O'Reilly Media.
[18] http://www.isaca.org/COBIT/
[19] http://www.itil-officialsite.com/
[20] http://www.dependability.org/wg10.4/
[21] Avizienis, A., J.-C. Laprie, B. Randell and C.E. Landwehr. 2004. Basic Concepts and Taxonomy of Dependable and Secure Computing, IEEE Trans. On Dependable and Secure Computing, Vol. 1, No. 1, Jan.–March 2004, pp. 11–33.
[22] Kano, T. and Y. Kikuchi. 2006. The Dependable IT Network, NEC Technology, Vol. 59, No. 3, pp. 6–10.
[23] Matsuda, K. 2009. Foreword, IPA SEC Journal No. 16, Vol. 5, No. 1 (Issue 16), p. 1.
[24] Descartes, R. (translated by Tanikawa, Takako). 1997. Discourse on the Method, Iwanami Shoten.
[25] Bertalanffy, L. 1969. General System Theory: Foundations, Development, Applications, G. Braziller, New York.
[26] Tanenbaum, A.S. and M. Van Steen. 2006. Distributed Systems: Principles and Paradigms, Second Edition, Prentice Hall.
[27] Bernstein, P.A., V. Hadzilacos and N. Goodman. 1987. Concurrency Control and Recovery in Database Systems. Addison-Wesley.

3

THE **DEOS** TECHNOLOGICAL SYSTEM

In the previous chapter, we showed how approaches to dependability have changed over the years, and went on to define Open System Dependability (OSD) as a new dependability concept and discuss the basic methods for making this concept a reality. In order to satisfy the requirements of OSD, a system must possess functionality for accommodating changes in its objectives and its environment, functionality that facilitates continuous operation, and functionality for ensuring that accountability with regard to the system can be consistently achieved; in addition, this functionality must be integrated into the system itself in the form of a repetitive process. The dependability of the system can thus be improved in iterative steps. In this chapter, we will lay out the structure of the specific technologies required to realize OSD as the Dependability Engineering for Open Systems (DEOS) Technological System.

We have shown that the achievement of OSD requires functionality for accommodating changes in the objectives and environment of the system. This *change accommodation* functionality relates to system changes that occur both during development and after the start of operation, and it is closely tied to the process of system development. We have also shown that, in order for the system to deliver the expected services to users as continuously and safely as possible, it must have both *failure prevention* functionality and *responsive action* functionality. These preventive action and failure response functions are linked to the process of system operation. Whenever the failure preventive and responsive action functionality have been utilized, changes must be made to the system in order to realize *recurrence prevention*. That functionality initiates system development from the system operation stage.

To ensure that accountability can be consistently achieved, analysis of the causes of a failure *(cause analysis)* must be performed promptly. To do this, the system needs (1) *agreement log retention* functionality for structured description of the agreements reached by stakeholders with regard to system requirements and their implementation, and (2) *operation log retention* functionality for monitoring and recording the operational status of the system. Accountability refers to the ability to demonstrate at any time either that system development and operation were correctly implemented or that, as a result of certain factors, this was not the case. This property is linked to both the system development and system operation processes. Actually, each stakeholders' agreement is associated with either system development or system operation, therefore, agreement log retention is linked to both. Operation log retention functionality is also linked to both development and operation: it is utilized during system operation, but what and when to monitor and record depend on the agreements reached in advance by stakeholders.

In light of the above, if a system is to continually provide the expected benefits to its users while accommodating changes in requirements and ensuring that accountability can be consistently achieved, the *DEOS Process* must be an iterative process that integrates both the *development cycle* and the *operation cycle*. In addition, that integrated, iterative *DEOS Process* must include recurrence prevention functionality, which initiates system change in response to operation of the failure prevention functionality or responsive action functionality. We define such a process as the *DEOS Process*, details of which will be described below.

First of all, we will define the *Change Accommodation Cycle*, the *Failure Response Cycle*, and the *Ordinary Operation state* from the perspective of dependability. The development process as it has existed to date corresponds to the *Change Accommodation Cycle*; the operation process, to the *Ordinary Operation state* combined with the *Failure Response Cycle*. In the *DEOS Process*, the *Change Accommodation Cycle* and the *Failure Response Cycle* are configured as a double loop that is initiated from the *Ordinary Operation state*. As described below, the *Change Accommodation Cycle* comprises a Consensus Building process, a Development process, and an Accountability Achievement process, while the *Failure Response Cycle* includes a Failure Response process and the Accountability Achievement process. Thus the *DEOS process* is a "process of processes".

We have seen that accountability achievement requires both structural description of stakeholders' agreements about system requirements and their implementation, and functionality for maintaining logs of those agreements, but how exactly should the stakeholders' agreements be described in notation, and how should the logs be maintained? The stakeholders must agree on those requirements that concern the design, development, and operation of the system and the methods that will be used to satisfy those requirements. Matters that require agreement must be argued using a suitable approach, and evidence that supports the argument needs to be presented so the stakeholders can themselves be fully assured and they can assure others. We therefore developed the D-Case by extending the assurance case [1] in order to provide a structured notation method for describing a logical consensus-building structure together with the required assumptions and evidence. We also developed the DEOS Agreement Description Database (D-ADD) as a means for storing D-Case logs and tracking them to support accountability.

The second functionality required in order to achieve accountability is *operation log retention*. For this purpose, we have defined a DEOS Runtime Environment (D-RE), which is analogous to a computer's operating system. D-RE implements functionality for monitoring and recording of system operation on a kernel for secure execution, and it can be used to execute the Failure Prevention and Responsive Action functions, as well as to provide information for analyzing the causes of a failure. Instructions for system monitoring and recording, in addition to the execution of the Failure Prevention and Responsive Action processes, must be based on the advance agreement of stakeholders; accordingly, we have provided the required description functionality in the form of D-Case, and we have developed D-Script as a secure script for giving instructions to the various functions of D-RE. In addition, D-RE includes a D-Script Engine for executing D-Scripts.

Achieving accountability using these functions forms a part of each of the *Change Accommodation Cycle* and the *Failure Response Cycle*, and it is incorporated into the *DEOS process* as a whole. Yet if this function set is not configured as an architecture together with other tools, it will not be possible to implement the *DEOS Process*. Accordingly, the *DEOS Architecture* comprises D-ADD (which stores D-Cases and D-Scripts) and D-RE (which provides the monitoring and recording, failure prevention, and responsive action functions), as well as tools for requirements elicitation and risk analysis, stakeholders' agreement tools, and other tools for application development among other components.

We discuss the *DEOS Process*, D-Case, and the *DEOS Architecture* in detail in the following sections. The D-Case description that assures the execution of DEOS is then described.

3.1 THE DEOS PROCESS

As shown in Fig. 3-1, the *DEOS process* is configured as follows:

1. The process consists of two cycles—the *Change Accommodation Cycle* (outer loop) and the *Failure Response Cycle* (inner loop)—both of which are initiated from the *Ordinary Operation state*.

2. The *Change Accommodation Cycle* begins when the system is to be modified in response to changes in its objectives or environment, and it comprises the Consensus Building process, the Development process, and the Accountability Achievement process.

3. The *Failure Response Cycle* begins when a failure has occurred or is predicted, and it consists of the Failure Response process and the Accountability Achievement process.

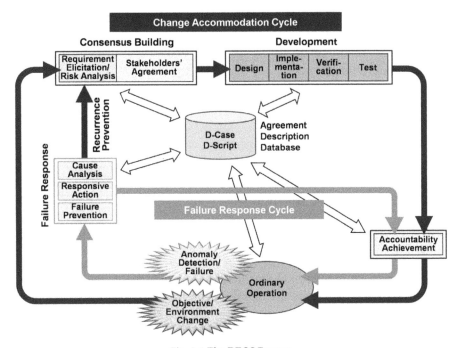

Fig. 3-1 The DEOS Process.

4. If deemed necessary after the cause of the failure has been analyzed, the *Failure Response Cycle* can initiate the *Change Accommodation Cycle* in order to modify the system.

5. A D-Case is a structured description of agreements reached by stakeholders concerning the requirements placed on the system and the methods to be used to satisfy them.

6. D-Scripts express in the form of scripts the instructions for monitoring and recording system states and procedures for recovery of the system from failures, both of which are agreed upon by the stakeholders.

7. The agreement description database, which is known as D-ADD, stores D-Cases and D-Scripts on an ongoing basis.

We will now describe the *DEOS Process* in detail.

3.1.1 Stakeholders

We have not yet presented a definition of stakeholders in regard to the DEOS Technological System, so we will do so now: A stakeholder of a system is any party concerned with the dependability of that system. We have identified the following as potential stakeholders:

- Users of system services or products (i.e., customers, or society at large in regard to social infrastructure systems)
- Providers of system services or products (usually business owners)
- System providers
  - Designers and developers
  - Maintenance personnel
  - Hardware providers
- Approval and authorization agencies

Within the *DEOS Process*, stakeholders explicitly argue for system requirements from their own standpoints. When consensus is reached concerning requirements and how they will be satisfied, the details of the agreement are recorded so that development, modification, and operation will be initiated accordingly. This process occurs continuously over the lifecycle of the system.

The stakeholders' requirements will change over time for many different reasons, which are likely to include the need to change the makeup of services in order to vie with competitors; substantial customer demands for new services; system modifications due to a merger or acquisition; technological advancements that make it possible to achieve the same functionality at a cheaper price; and compliance with new laws or standards. The longer a system is in operation, the more its requirements will evolve, and normally, it is the stakeholders who decide when action should be taken in response. In the *DEOS Process*, we identify this type of requirement change as being due to *changes of objectives or environment*.

System modifications needed to prevent a failure from reoccurring must also be carried out in accordance with the stakeholders' agreements. In contrast to changes in objectives or environment, however, rapid response is often required for this type of change.

3.1.2 Ordinary Operation

The term "ordinary operation" is used to describe the state of the system when it is providing normal services to users in an ongoing manner without deviating from the allowable range of service levels agreed upon by the stakeholders. This service range is referred to as the *In-Operation Range* (IOR). The most important function executed in the *Ordinary Operation state* is detection of signs of potential failure or the actual occurrence of failure. In this state, therefore, the monitoring functionality provided by D-RE is used to observe important system parameters, and the system must be able to determine whether the values of these parameters are outside the IOR. Deviation from the IOR is detected as a failure. It is also possible to detect signs of potential failure from the pattern of change in the system state even when all parameters are inside the IOR. If a failure or a sign of potential failure is detected, the *Failure Response*

Cycle is initiated from Ordinary Operation. Factors such as the parameters to be monitored, the frequency of monitoring, how to process the results obtained, and how to identify actual failure or signs of potential failure are agreed upon in advance by the stakeholders. These agreements are described using D-Cases, and they are put into effect using D-Scripts generated based on the D-Cases.

Another important function for Ordinary Operation is the detection of changes in objectives or environment. Given that these types of change occur outside the system, it is difficult to automate detection. Still, this functionality can be realized by creating a means for constantly monitoring business objectives, the behavior of users, technology trends, changes in regulations or standards, and so forth, by assigning the necessary personnel, and by also establishing rules concerning how these personnel should report any such changes at stakeholder meetings and other suitable venues. It is also crucial that rules be put in place for regular review of the system's objectives and environment. When any change in the objectives or environment is confirmed, the *Change Accommodation Cycle* is initiated.

In addition to the above-described processes, a number of background processes are executed in the *Ordinary Operation state* for routine inspection of operation logs, regular review and improvement of processes, training of personnel, and so on. Keeping the system's memory resources in a clean condition is a highly effective element of routine maintenance and improvement work. Meanwhile, simulation can be performed with the system time advanced in order to detect signs of potential failure before they would normally become apparent.

3.1.3 Change Accommodation Cycle

The *Change Accommodation Cycle* is implemented in order to adapt the system to any changes in stakeholder objectives or the external environment. The main processes that make up this cycle are the Consensus Building process (comprising the Requirement Elicitation & Risk Analysis phase and Stakeholders' Agreement phase for dealing with system change), the Development process (comprising the Design, Implementation, Verification, and Testing phases), and the Accountability Achievement process. The *Change Accommodation Cycle* is also initiated when, as a result of execution of the Cause Analysis phase in the *Failure Response Cycle*, system changes must be implemented in order to prevent reoccurrence of a specific failure or occurrence of a similar one.

The Requirement Elicitation & Risk Analysis phase is initiated first of all when a change in the system's objectives or environment has led to a change in the requirements of the stakeholders (or the addition of new requirements) or when cause analysis in response to a failure indicates that the system must be changed. In each of these cases, a set of functional requirements for the system based primarily on the service objectives of the business owner but also taking into consideration the needs of users, the system environment, advances in technology, and relevant regulations and standards are elicited in this phase. Service continuity scenarios are also created for the system at this time on the basis of the service objectives. Risk is analyzed using these scenarios, and the result is a set of non-functional requirements, including dependability requirements.

In the Stakeholders' Agreement phase, the stakeholders negotiate regarding the dependability-related requirements of the system and the methods to be used to realize them. This is done based on the requirements elicited as described above, and the details of the resulting agreements are described in the form of D-Cases. Also, based on the service continuity scenarios, the stakeholders create D-Scripts describing the corresponding executable procedures. Together, the Requirement Elicitation & Risk Analysis phase and the Stakeholders' Agreement phase form the Consensus Building process.

The phases of Design, Implementation, Verification, and Testing constitute the Development process. Much research has been done in the field of development processes, and many software processes and tools have been developed. The best of these should be utilized in a proactive manner. As part of the DEOS Project, we have also developed a set of tools for strengthening the *DEOS Process*, such as programs for software verification [2], benchmarking [3], and fault injection testing [4].

The Accountability Achievement process requires the service or product providers to explain the background, nature, and timing of any alterations of services and functions whenever the system is modified to satisfy changes in stakeholder requirements related to the objectives and environment. This process is also used when there is a need for explanation concerning either the status of regular services or aspects of design, development, maintenance, and operation. As such, the Accountability Achievement process is crucial for maintaining the reputation of the service or product provider among individual users and society at large, in addition to ensuring the profitability of its business. Agreement logs stored in the form of D-Case description in combination with operation logs, both of which are retained in the agreement description database, play an important role in achieving accountability.

Ideally, the *Change Accommodation Cycle* should run in parallel with *Ordinary Operation* so that the system can be changed without interrupting the provision of services.

3.1.4 Failure Response Cycle

When a failure or a sign of a potential failure has been detected in the *Ordinary Operation state*, the *Failure Response Cycle* acts immediately to minimize damage or to prevent the failure from occurring. In terms of the *DEOS Process*, a *failure* is defined as any deviation from the IOR—that is, the allowable range of service and function levels agreed upon by the stakeholders.

The *Failure Response Cycle* comprises the Failure Response process, which is made up of the Failure Prevention, Responsive Action, and Cause Analysis phases, and the Accountability Achievement process required in the event of an actual failure. The Failure Prevention, Responsive Action, and Cause Analysis phases do not need to be mutually independent or executed in any specific order. They are often implemented with a closely inter-related configuration in order to operate as an integrated whole.

Whenever failure is predicted during operation of the system or an increase in the probability of a failure has been detected, the Failure Prevention phase takes action to prevent actual occurrence. It can take effective countermeasures when the prediction is sufficiently early. For example, additional resources could be allocated to the system or throughput could be reduced by limiting the current resources; alternatively, rejuvenation could be employed to prevent the system going offline, or in the worst case scenario, time could be gained before this actually occurs. If a failure cannot be predicted until just before it occurs, the aim of the Failure Prevention phase is to minimize its effect. It can also record internal system information between prediction and occurrence of a failure—information that can then be put to effective use in identifying the cause. One means of predicting failure would be to analyze past failure patterns in order to identify when similar ones reoccur. Failure prevention scenarios must be agreed upon by the stakeholders and described in D-Scripts in advance. Preventive actions can then be executed either automatically or with the cooperation of operators and the system administrator.

The purpose of the Responsive Action phase is to minimize the effect of any failure. The first step is to determine what measures must be implemented right away and execute the corresponding processes. Normally, the Responsive Action phase isolates the failure in order to localize its effect and prevent all services from going offline. This involves terminating operation of the application or system component where the failure occurred and performing a reset. Following this, the operator or system administrator restores normal operation. Scenarios for rapid failure response must be described in D-Scripts in advance based on stakeholders' agreements, and ideally should be executed automatically. That said, the Responsive Action phase must also deal with unforeseeable failures. In preparation for such a situation, emergency response plans for service continuity (detailing the responsible parties or groups, procedures, escalation paths, and the like) must be established in advance in line with the objectives of each service, and the stakeholders must agree on the corresponding procedures. The effect of any failure can then be quickly minimized based on the instructions provided in the plans through the cooperation of operators and the system administrator.

In the Cause Analysis phase, the reason for a failure is identified from operation logs and agreement logs in the form of D-Case, both of which are stored in D-ADD. The structure of arguments made in reaching consensus as well as the corresponding assumptions and evidence are described in D-Cases, and therefore, working through the agreement logs can yield extremely useful information for identifying causes of failure, such as differences in how the meanings of assumptions have been interpreted, changes in the assumptions due to the passage of time, incomplete arguments, and omissions or errors in the scope of evidence in the form of fault-tree analysis, benchmark testing results, and so forth. The *Change Accommodation Cycle* is then initiated to ensure that an identical or similar failure will not occur again in the future.

In the Accountability Achievement process initiated whenever a failure occurs, service and product providers must provide the users of their services and products as well as all other stakeholders with a satisfactory explanation concerning the status, cause, and magnitude of the failure in addition to other relevant matters such as planned countermeasures and steps already taken in response. On occasion, they may also be required to explain compensation amounts and how responsibility will be apportioned. Ensuring that a system is accountable is of considerable importance in maintaining a good reputation with both individual users and society at large, and also in terms of the profitability of the service provider's operations.

As with the *Change Accommodation Cycle*, the *Failure Response Cycle* should ideally be executed during Ordinary Operation. In practical terms, it may be possible to keep services online when the system detects signs of a potential failure by automatically executing avoidance processes within the allowable range of service and function levels described in the D-Scripts. Alternatively, certain functions could be operated at a reduced level to maintain services overall. That said, the complete cessation of services may be unavoidable in certain cases.

3.2 D-CASE AND D-SCRIPT

3.2.1 D-Case

D-Case is a structured notation method for describing the agreements of stakeholders concerning system requirements. The description created based on this method is also referred to as a D-Case. Stakeholders' agreements are assured only when their assumptions are neither excessive nor insufficient, when they take the form of sound arguments, and when they are supported by suitable evidence. In the following paragraphs, we will describe this property of assuredness and introduce D-Case. A more detailed description of D-Case can be found in Chapter 4.

As stated previously, it is impossible to give a complete description of an open system. Such a description only exists when it contains no contradiction, all aspects of the system are defined, and all concerned parties interpret those definitions in the same way. If this is impossible, then what is our best alternative? We believe that it is to make the description of the system as complete as possible. From the standpoint of practical implementation of a system, the validity of its description must be contingent on all concerned parties (in DEOS terms, the stakeholders) sharing a common understanding of all relevant matters. That is to say, we can move closer to a complete description of a system through consensus building that ensures the mutual assurance of all stakeholders.

Assuredness can be achieved by fully describing assumptions, arguing in an appropriate manner, and presenting suitable evidence. It serves as the basis of logical proof and is also a means of presenting legal arguments in a courtroom. The general form of this method is known as the *assurance case*. Based on this approach, the safety case and the dependability case [1] have been proposed as methods for developing assured safety and dependability arguments, and they are starting to be put to practical use.

While assuredness unfortunately does not guarantee completeness, it does enhance the credibility of an argument. In terms of the DEOS Technological System, assuredness improves the reliability of satisfying requirements [5]. For these reasons, assuredness has been adopted as the basis for DEOS consensus building, and the assurance case method is used for agreement description.

When developing an assurance case

- The fact to be assured (claim) is presented as the *goal*.
- Conditions relevant to the goal are described as *contexts*.
- Goals can be successively decomposed into sub-goals—the goal being supported only when all of its sub-goals are supported.
- Sub-goals are supported by suitable *evidence*.
- Evidence can be provided as documentation of design, implementation, verification, and testing methods or results, or as operation log records and the like.

The assurance case is a very powerful form of description of agreements concerning development. However, development and operation cannot be isolated from one another when assurance of the dependability of an open system is the ultimate aim. In order, therefore, to make the assurance case applicable within the framework of the DEOS Technological System, we have added *monitoring* and *action* as new evidence categories. A monitor node issues a command to monitor a certain system parameter and returns its values in real time. It can then be determined whether or not this value is within the IOR. Action nodes are needed to issue commands to isolate, abort, or reboot a process in which a failure has occurred, and these commands are executed using D-Scripts. We have also added an *external* node in order to link to software modules that were either developed or exist outside the system. This node can reference the D-Case notation of external systems, and as such, it provides a means for dependable implementation of software developed by third parties and external services.

Goal Structuring Notation (GSN) [6] with goal, strategy, context, evidence (including monitoring and action), external, and undeveloped nodes has been adopted as the main form of notation for developing D-Cases. The descriptions within these nodes are usually written in natural language. Ambiguity should be removed as much as possible, and the use of pseudo-natural languages such as SBVR and Agda is recommended as a means of facilitating automatic consistency checking.

Figure 3-2 shows an example of an agreement in D-Case notation. Here, a goal—namely, that the system can tolerate failure due to response-time delays—is described in *Goal: G_1*. Linked to this goal is *Context: C_1* which describes an assumption—that the IOR for response times is Normal at less than 50 ms, Severity Level 1 between 50 and 100 ms, and Severity Level 2 at over 100 ms. *Strategy: S_1* indicates that the goal in *G_1* is to be decomposed into sub-goals from the perspective of monitoring and recovery action. *Goal: G_2* claims that monitoring of the response time is possible, and the monitor node *Monitor: M_1* is linked as evidence for this sub-goal. In addition, *Goal: G_3* claims that recovery action is possible, *Context: C_2* describes the assumption that a D-Script has been prepared, and *Evidence: E_1* is linked to provide test results as supporting evidence.

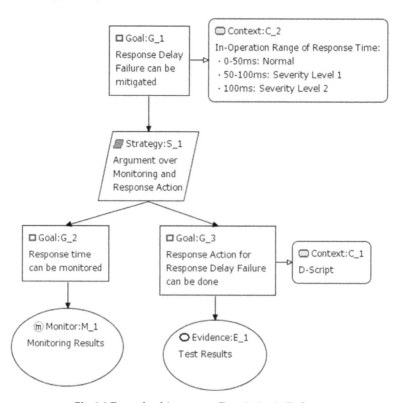

Fig. 3-2 Example of Agreement Description in D-Case.

3.2.2 D-Script

D-Case is not limited to agreements concerning system development: it can also describe operation-related agreements. For example, it can be used for agreements concerning monitoring and recording or the detection of failure or signs of potential failure in the *Ordinary Operation state*, as well as preventive action, responsive action, and cause analysis within the *Failure Response Cycle*. Details such as what to monitor, what data to retain in logs, and what basis to use for identifying failure or signs of failure must, on the basis of service continuity scenarios, be agreed in advance by the stakeholders and described as D-Cases. Similarly, the stakeholders must agree in advance on what specific processes to execute for preventive action, responsive action, and cause analysis upon detection of a failure or a sign of a potential failure. D-Script has been introduced into the *DEOS Process* to provide for flexible failure management based on such descriptions.

As already mentioned, D-Case includes monitoring and action nodes that can be used to describe arguments concerning the *Ordinary Operation state* and the *Failure Response Cycle*. That is to say, the stakeholders must reach agreement on the basis of service continuity scenarios specifying what system conditions to monitor and what values will trigger what sorts of responses. Such an agreement is described in the form of a D-Case, which is then used to generate D-Scripts. In some situations, D-Scripts can be derived automatically from D-Case description, but many involve value judgments concerning risk, meaning they must be created manually. Accordingly, the D-Scripts themselves are also subject to stakeholders' agreement.

The following section illustrates how, in order to execute the actions described in D-Scripts, D-RE has been provided with a D-Script Engine as well as functionality for execution control in line with D-Script instructions. D-Scripts created on the basis of D-Cases, the D-Script Engine for execution of those scripts, and D-RE's related functionality make it possible to achieve failure response as agreed by the stakeholders in the actual system.

3.3 DEOS ARCHITECTURE

The *DEOS Process* allows Open Systems Dependability to be achieved in an iterative fashion. In order to apply this approach to a real-world system, however, the system under consideration must have a number of basic functions. Specifically, execution of the *DEOS Process* requires the DEOS Runtime Environment (D-RE), and this environment—which is analogous to a computer's operating system—must have functionality for monitoring, recording of data, failure isolation, and execution of D-Scripts. In addition, a D-ADD is needed to store agreement logs (in D-Case and D-Script) and operation logs. Consensus building tools are also required, as is functionality for software verification and benchmarking. In terms of the

DEOS Technological System, these and all other component elements of the overall structure for employing the *DEOS Process* are collectively referred to as the *DEOS Architecture*.

The actual implementation of the *DEOS Architecture* may differ depending on the level of dependability required in the system under consideration. Below, however, we will describe the fundamental architecture developed on the assumption of application to the massive, highly complex information systems in operation today. Looking at the *DEOS Process* and the *DEOS Architecture* side by side, it is easy to understand how the former is executed within the actual system.

The component elements of the *DEOS Architecture* are as follows (Fig. 3-3):

- A set of support tools for each of the Requirement Elicitation and Risk Analysis phase and the Stakeholders' Agreement phase.

- DEOS Development Support Tools (D-DST), which include program verification tools as well as a set of tools for benchmarking and fault injection testing.

- The DEOS Agreement Description Database (D-ADD) containing agreement logs in D-Case, D-Script, and operation logs as the results of D-Script execution.

- The DEOS Runtime Environment (D-RE).

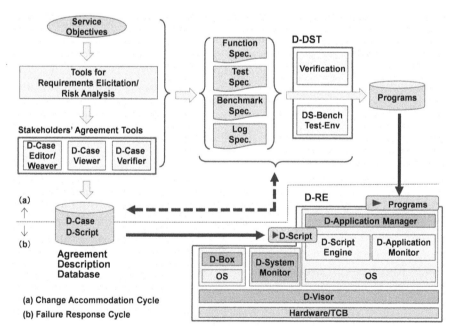

Fig. 3-3 The DEOS Architecture.

3.3.1 Consensus Building Tools

In the Requirement Elicitation & Risk Analysis phase, the system's functional requirements are identified based on the service objectives of the business owner, while also taking into consideration the needs of the user, the system environment, and relevant regulations and international standards. Also in this phase, service continuity scenarios are created for foreseeable types of failure, risk analysis is carried out, and the non-functional requirements, which include dependability requirements, are elicited.

In the Stakeholders' Agreement phase, agreements are described in D-Cases in line with consensus building methods and description methods. Tools for this purpose include D-Case Editor, which was created primarily using Eclipse, as well as D-Case Weaver—a JavaScript application for use in web browsers.

3.3.2 DEOS Development Support Tools

The DEOS Development Support Tools (D-DST) provide support for the design, development, verification, benchmarking, and testing of programs on the basis of functional specifications, test specifications, benchmarking scenarios, and log specifications that have been developed to match business objectives and service continuity scenarios. A wide range of other development support tools are available on the market, and these can be applied as appropriate. Developed as part of the DEOS Project, DS-Bench/Test-Env is an original tool for software verification using type checking and model checking, as well as benchmarking with support for fault injection.

3.3.3 DEOS Agreement Description Database

The purpose of the DEOS Agreement Description Database (D-ADD) is to store dependability-related information consisting of agreement logs such as D-Case logs, D-Script logs, and documents used for requirement management, and operation logs as the results of D-Script execution, so that they can be efficiently accessed whenever necessary. Specifically, the following materials are stored within D-ADD:

- Fundamental documents describing the basic structure and basic components of the system under consideration;
- D-Case and D-Script description logs;
- Any documents that constitute D-Case evidence;
- Operation logs as the results of D-Scripts execution; and
- Any information related to past failures and signs of potential failure, as well as information concerning countermeasures taken and the results.

D-ADD plays a central role in regard to all elements of the *DEOS Process*. It employs a three-layer structure, comprising a Fundamental Tools layer, a Core layer, and a Hybrid Database layer (Fig. 3-4). The Fundamental Tools layer provides interfaces and tools; the Core layer provides structural models for recording of evidence and system status; and the Hybrid Database layer is used to physically record and store the corresponding information.

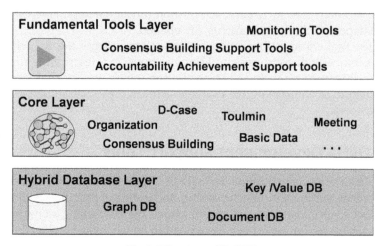

Fig. 3-4 Structure of D-ADD.

3.3.4 DEOS Runtime Environment

The purpose of the DEOS Runtime Environment (D-RE) is to ensure that dependable services can be delivered. It comprises the following sub-systems.

- *D-Visor*: A virtualization layer providing system containers that secure the independence of the component elements so the system can be reconfigured. For example, an error or failure occurring in one specific system container can be prevented from propagating to any others.

- *D-System Monitor*: This sub-system provides functionality for monitoring the system.

- *D-Application Manager*: A component of D-RE providing application containers that secure the independence of multiple applications; in addition, it also manages and controls the lifecycle (launch, update, and shutdown) of each application.

- *D-Application Monitor*: This sub-system provides functionality for monitoring the operation of applications as well as gathering evidence and storing it in a D-Box.

- *D-Box*: A storage area that is used to safely and securely record evidence and other related information.

- *D-Script Engine*: The role of this component is to execute D-Scripts safely and reliably, and it controls D-Application Manager, D-Application Monitor, and D-System Monitor in order to do so.

3.4 ASSURANCE OF DEOS PROCESS EXECUTION USING D-CASE

Describing the DEOS Process with D-Case makes it possible to assure execution of this process in line with the relevant agreements. D-Case also provides monitoring and action functionality for checking whether the system is operating based on the content of agreements and for controlling this operation. In this section, we will first describe the fundamental parts of the *DEOS Process* using D-Case and assure execution of the process. We will then look at how the D-Case monitoring and action functions described in D-Scripts operate based on the D-Case agreements.

3.4.1 Describing the Basic DEOS Structure

As we have seen, the *DEOS Process* comprises three basic elements—namely, the *Ordinary Operation state*, the *Change Accommodation Cycle*, and the *Failure Response Cycle*. Explicitly defining these components is a fundamental feature of DEOS. D-Case allows a wide range of assurance targets to be described from various perspectives and with a range of different policies; however, we should describe D-Case for an assurance target from the perspective of the DEOS basic structure in order to assure execution of the *DEOS Process*. For this, the D-Case top goal should be decomposed according to the above-mentioned three elements.

Figure 3-5 illustrates this decomposition in D-Case. Here, the top goal is described as "Service continuity and accountability can be achieved in an ever-changing system", and this is divided into three sub-goals according to the *DEOS Process* elements.

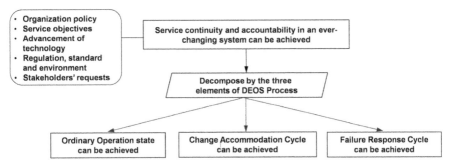

Fig. 3-5 Top Goal of DEOS Basic Structure in D-Case.

In order to develop the sub-goal "*Ordinary Operation state* can be achieved", evidence for the functionality of change monitoring and failure monitoring is defined with operation rules, daily inspection guides, and the like as context, and the results obtained from this evidence are evaluated (Fig. 3-6).

In order to provide evidence to support the sub-goal "*Change Accommodation Cycle* can be achieved", it is crucial that procedural documents for change accommodation and for accountability achievement have been produced for the procedures to be integrally executed upon the detection of any change in the objectives or environment of the system (Fig. 3-7).

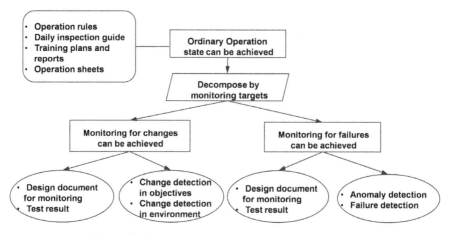

Fig. 3-6 D-Case Description for Ordinary Operation State.

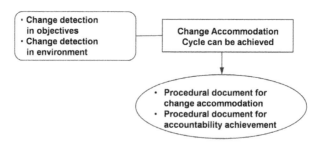

Fig. 3-7 D-Case Description for Change Accommodation Cycle.

The sub-goal "*Failure Response Cycle* can be achieved" is decomposed according to whether failures are foreseeable or unforeseeable. In reality, the only conditions that can be argued are foreseeable ones; however, in order to demonstrate that unforeseeable failures are also possible, we use "unforeseeable failure" here where "other failure" would normally be applied. Meanwhile, countermeasures cannot be devised for unforeseeable failures, so we instead define how individuals or the organization as a whole should respond to such an event (Fig. 3-8).

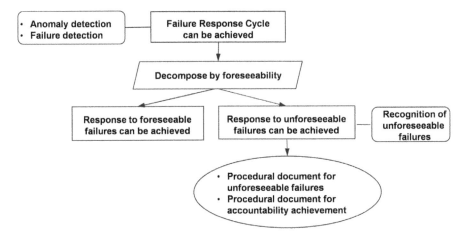

Fig. 3-8 D-Case Description for Failure Response Cycle.

The sub-goal "Response to foreseeable failure can be achieved" is decomposed according to whether or not the failure was planned for at the design stage. Here, the term "response to failure that was planned for in design" refers to the action required of the system in response to all conceivable types of failure; in other words, it corresponds to the development of service continuity scenarios. This entails the production of specifications and design documents, implementation, and testing, and all of this constitutes D-Case evidence. Certain failures may not have been planned for in design due, for example, to cost-related factors, and the action to be taken in response to such a failure by individuals or the organization as a whole must be decided (Fig. 3-9).

Fig. 3-9 D-Case Description for Foreseeable Failure.

Figure 3-10 shows the entire D-Case for the basic DEOS structure. This D-Case description forms a standard pattern in describing D-Cases for various systems, and it is called the *DEOS Process Decomposition Pattern* in Section 4.6. The D-Case for a real-world system could be produced by further expanding and refining this description.

The broken-line arrows in Fig. 3-10 show how individual evidence results serve as contexts for sub-goals to the right. This means that if a phenomenon at the base of any of these arrows were to occur during actual operation, the context at the tip of the arrow would come into effect and processes for supporting the corresponding goal would be initiated.

In the event of a requirement to accommodate change, D-Case for the next version of the system is developed from the D-Case of the current version (Fig. 3-11). Storing of agreement and operation logs in relation to system updates of this nature can be extremely valuable in terms of achieving accountability in the *DEOS Process*. As a result, the system's long-term running costs can be reduced and the service provider's profit-making opportunities secured; in addition, the service provider's brand and reputation with customers can be protected.

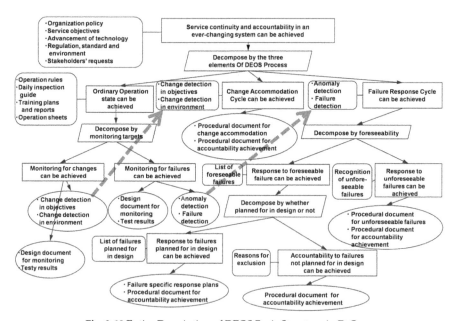

Fig. 3-10 Entire Description of DEOS Basic Structure in D-Case.

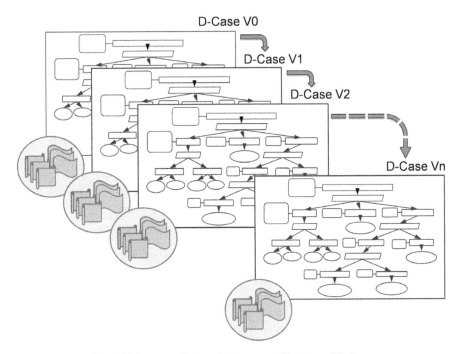

Fig. 3-11 Accommodation of Change and Revision of D-Case.

3.4.2 Controlling System Operation based on D-Case

Under the DEOS approach, system operation is described using monitoring and action nodes within D-Cases. Monitor nodes indicate the targets, timings, and methods of monitoring and recording. Action nodes indicate how the system behaves. Descriptions of operation-related agreements based on these nodes are represented as scenarios using D-Scripts, and these scripts are executed by the D-Script Engine within D-RE.

Inside D-RE, the D-System Monitor and D-Application Monitor provide operating-system monitoring functionality and application-program monitoring functionality, respectively. Monitoring by the monitor nodes is supported by these subsystems. As mentioned previously, D-Visor provides system containers and D-Application Manager provides application containers. Action nodes use this functionality to isolate, abort, or reboot processes in which failures have occurred.

The DEOS approach requires that all aspects of development and operation be agreed upon by the stakeholders, described using D-Cases, and utilized during program development, while also being executed as D-Scripts during operation. Thus D-Case ensures that all stages from development through operation are executed on the basis of the stakeholders' agreements.

REFERENCES

[1] Bloomfield, R. and P. Bishop. 2010. Safety and Assurance Cases: Past, Present and Possible Future—an Adelard Perspective, Proceedings of the Eighteenth Safety-Critical Systems Symposium, Bristol, UK, 9–11 February 2010, pp. 51–67.

[2] Matsuda, M., T. Maeda and A. Yonezawa. 2009. Towards Design and Implementation of Model Checker for System Software. In Proc. of First International Workshop on Software Technologies for Future Dependable Distributed Systems (STFSSD), pp. 117–121.

[3] Fujita, H., Y. Matsuno, T. Hanawa, M. Sato, S. Kato and Y. Ishikawa. 2012. DS-Bench Toolset: Tools for dependability benchmarking with simulation and assurance, IEEE/IFIP Int'l Conf. on dependable Systems and Networks (DSN 2012).

[4] Hanawa, T., H. Kiozumi, T. Banzai, M. Sato and S. Miura. 2010. Customizing Virtual Machine with Fault Injector by Integrating with SpecC Device Model for a software testing environment D-Cloud, In Proc. of the 16th IEEE Pacific Rim International Symposium on Dependable Computing (PRDC'10), pp. 47–54.

[5] Object Management Group Standard. 2013. Structured Assurance Case Metamodel (SACM), Version 1.0, OMG Document number: formal/2013-02-01, Standard document URL: http://www.omg.org/spec/SACM/

[6] The GSN Working Group. 2011. GSN Community Standard, Version 1.

D-CASE—BUILDING CONSENSUS AND ACHIEVING ACCOUNTABILITY

4.1 CONSENSUS BUILDING AND ACCOUNTABILITY

As pointed out in IEC 61508 and ISO 26262, the safety of systems must be demonstrated to the satisfaction of society at large. In order to prove that a system is safe, it must be shown to pose no threat to the people or the environment. The same applies to dependability—owners must demonstrate that their systems satisfy all dependability requirements under normal operation conditions.

Specifically, the dependability of a system must be argued in order to support each of the following three claims:

1. A failure will not occur in the system;
2. Even if a failure were to occur in the system, appropriate action could be taken to ensure continuity of business; and
3. The system is capable of identifying, recovering from, and preventing reoccurrence of an unforeseen failure.

In cases 1 and 2, we must present society at large with satisfactory and understandable reasons why, respectively, failure cannot occur in the system under the assumed conditions and the system is able to take suitable failure countermeasures. For case 3, proof that the cause of any such failure can be rapidly identified and reoccurrence prevented must be presented to the satisfaction of society. Here, providing proof of reoccurrence countermeasures is equivalent to demonstrating that a similar system failure cannot occur a second time, which is essentially the same as the argument made for case 1. In each of these cases, it will be difficult to ensure continuity of business using the system under consideration unless society can agree that an adequate level of accountability has been achieved.

When aiming to build consensus regarding the dependability of a system, it is important to identify the stakeholders to whom the argument must be made, to determine accurately the scope of the required argument, and to ensure a high level of precision. If important stakeholders are overlooked, rework will become necessary when their importance is recognized later on. It would be acceptable if this were to happen during consensus building, but recognition of important stakeholders would be a serious matter if it were not to occur until working to achieve accountability. Likewise, omission of any critical fact when arguing that a system is dependable suggests that the comprehensive nature of dependability itself has not been fully understood and that failure in the corresponding system component could potentially be overlooked.

We can employ a product approach or a process approach in order to determine the scope of the required argument. With the product approach, we must clearly identify the extent to which the dependability of a product must be argued in terms of its component elements. The process approach, meanwhile, typically focuses on (1) the development process, (2) the operation process, and (3) failure countermeasures and their effectiveness. With, for example, the well-known techniques of Fault Tree Analysis (FTA), Failure Mode and Effect Analysis (FMEA), and Hazard & Operability Studies (HAZOP), the aim is to identify the range of potential failures and the corresponding countermeasures. However, these approaches do not extend to verifying whether the countermeasures can be effectively implemented. They are insufficient in that they overlook the possibility that new, secondary failures could occur during the implementation of countermeasures. In accordance with the DEOS approach, therefore, the case for dependability must also demonstrate that *failure countermeasures can be implemented in practice and this will not result in secondary failure.*

In certain situations, it is not sufficient for dependability to be argued only on the basis of review and testing; instead formal methods, such as those employed with Evaluation Assurance Levels (EALs), must be used to achieve greater precision. It is important, therefore, to (1) identify explicitly the stakeholders who must agree with the case for system dependability and the scope of the desired agreement, and (2) clarify the basis for these decisions. The same applies to the DEOS approach.

When highly specialized notation is used in the pursuit of consensus regarding system dependability and accountability, it will be difficult to convince society at large. Standard notation methods must, therefore, be used for this purpose.

4.2 FROM ASSURANCE CASE TO D-CASE

4.2.1 Assurance Case

The safety case has become a common requirement under European and US safety certification standards for such high-safety systems as nuclear power plants. The assurance case is a concept derived from the safety case and generalized in order to include other characteristics such as dependability and security. One definition of "safety case" is:

> *A structured argument, supported by a body of evidence that provides a compelling, comprehensible and valid case that a system is safe for a given application in a given environment [1].*

The word "case" is used here with the generally accepted meaning of:

> *All the reasons that one side in a legal argument can give against the other side.* (Longman English Dictionary).

Meanwhile, "assurance" means both having confidence in some outcome and convincing others thereof. In the O-DA standard [13], which was standardized based on our work, "assuredness" is a key term. Although this standard does not clearly indicate who is to be assured, in a narrow sense, it is referring to certifiers of compliance with standards; in a broader sense, it can mean all stakeholders, including users, developers, and even society at large. A liberal interpretation of the meaning of "assurance case for dependability" is:

> *Arguments and documents for convincing stakeholders that the system under consideration is dependable.*

One of the main factors behind development of the safety case has been a significant shift in European and US safety certification standards since the 1970s, away from prescriptive regulations in favor of a goal-based approach. That shift was in large part a reaction to serious accidents, especially two which occurred in the UK in 1988—the Piper Alpha disaster that claimed the lives of 167 oil rig workers and the Clapham Junction rail disaster that killed 35. The term "safety case" was probably adopted as a result of its use in the report produced by Lord Cullen and his team in their analysis of the Piper Alpha disaster [2, 8].

Prescriptive methods for certification to standards require that certifiers confirm whether or not a set of prescribed requirements have been satisfied. In contrast, the goal-based approach sets forth qualities that a system should possess, such as safety and dependability. The system's developers and operators themselves make a case for why their system is safe using Fault Tree Analysis (FTA) and other similar techniques to provide evidence thereof. The documented arguments are then presented to the certifiers.

Bloomfield and Bishop have identified the following shortcomings with prescriptive regulations [3]:

- System developers and operators satisfy only mandated requirements to satisfy legal responsibilities. If it is later established that the mandated requirements were insufficient from the perspective of safety, responsibility lies with the regulators and regulations alone.
- Prescriptive requirements are mandated based on past experience, giving rise to a possible risk of these requirements becoming insufficient or unnecessary as the related technologies evolve.
- Prescriptive regulations can prevent system developers and operators from adopting new safety-related technologies.
- Prescriptive regulations can be barriers to open international markets and integration with other areas of science.
- Unnecessary costs are incurred in satisfying requirements mandated using the prescriptive approach.

In contrast, under the goal-based approach to certification, methods for satisfying a system's requirements are the responsibility of its developers and operators. With the option, therefore, to adopt freely any safety-related technologies considered suitable, they can use their best efforts to realize system safety. Bloomfield and Bishop have identified this as a significant advantage. Yet there is also strong criticism of the goal-based approach: for example, Leveson claims that most papers about safety cases express personal opinions or deal with how to prepare a safety case, but do not provide data on whether it is effective [4].

The typical structure of a safety case is shown in Fig. 4-1.

Some of the standards requiring the submission of safety cases are the Eurocontrol standards published by the European Organisation for the Safety of Air Navigation [5], the Yellow Book standards [6] for railways in the United Kingdom, and Defense Standard 00-56 [7] as adopted by the UK Ministry of Defense. In the United States, the medical device industry has recently been requiring the submission of safety cases, and they are already a requirement under the ISO 26262 functional safety standard for automobiles.

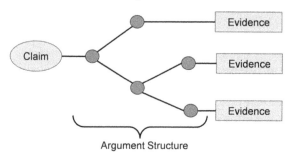

Fig. 4-1 Typical Structure of a Safety Case.

Safety cases are normally written in natural language, but few have been made publicly available because of the confidential nature of the corporate information that they contain. A number of examples of safety cases can be found in the Safety Case Repository maintained by the University of Virginia [9]. A range of graphical notation methods have been proposed to support description, interpretation, and communication of safety cases. Two of the most notable are Goal Structuring Notation (GSN) [10], which was developed at the University of York by Tim Kelly and other contributors, and Claims-Arguments-Evidence (CAE) [11], developed at Adelard by Robin Bloomfield and others. The D-Case notation that we developed (described below) is based on the GSN approach.

A typical example of CAE notation is shown in Fig. 4-2. Under this approach, cases are structured using three different types of node—claim, argument, and evidence. Claim nodes indicate requirements the system must satisfy. Argument nodes present arguments to support the claim. Evidence nodes present facts that ultimately support the claim.

A typical example of GSN is shown in Fig. 4-3. A goal is a proposition to be demonstrated. A context identifies assumptions and provides contextual information for arguing the goal. A strategy explains the method used to refine the goal and decompose it into sub-goals. Evidence is the ultimate footing for demonstrating the achievability of the goal. An undeveloped entity indicates that, at the present point in time, the goal is undeveloped in the argument or insufficient evidence exists to support it.

The claim, the evidence, and the argument of CAE correspond respectively to the goal, the evidence, and a combination of the goal and the strategy of

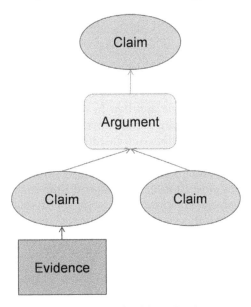

Fig. 4-2 Typical Example of CAE Notation.

Fig. 4-3 Typical Example of GSN.

GSN. In the basic sense, GSN and CAE are similar notation methods, and the Object Management Group (OMG) has standardized a combination of the two as the Structured Assurance Case Metamodel (SACM) [12]. It should be stressed that a safety case is not made up of a GSN or CAE description only, but also includes a range of other documents, models, and text. GSN and CAE editing tools must be linked to these other documents and to the corresponding model notation tools.

4.2.2 Definition and Origin of D-Case

D-Case is an extension of the conventional assurance case, adapted for open systems within the framework of the *DEOS Process* and serving as both a method and a tool. In the *DEOS Process*, the lifecycle of an open system comprises a range of phases that occur simultaneously, including development, operation, and failure response. Laprie et al. [14] and other researchers in the field of dependability have argued that the development and operation phases are clearly separate from one another. In order to facilitate argumentation using information from development, information from operation, and failure response details in a simultaneous fashion, we developed D-Case based on GSN, which has been used for system development, adding two new nodes—monitor and action (described below)—that extend the concept of evidence to include operation-monitoring information and failure-response details, which are used in operation. Thus, we realized D-Case for use both in development and operation including failure response. We also added a third extension—namely, an external node to describe inter-system relationships in an open system.

D-Case is defined as follows:

A method and tool to facilitate stakeholders' agreement on the dependability of an open system and to ensure accountability to society through all stages of the life of the system, from development through operation, maintenance, and decommissioning. The description in D-Case is also referred to as D-Case.

D-Case will be described in detail in the next section.

The safety case in its current form has mainly been applied for certification, although it is also being used by system development teams for communication among system providers and outside consultants. As information systems grow larger and more networked, many different stakeholders need to reach consensus throughout both the development and operation phases; furthermore, dependability must be achieved in a collaborative manner with the other systems. Bearing in mind the importance of integration with tools, runtime systems, and other elements of DEOS still being further improved, we determined the following three factors to be critical in terms of practical implementation:

- *Development of introductory documents and training courses that practitioners would find easy to comprehend*: In the past, safety cases for systems requiring a high level of safety have been produced by consultants with considerable specialist knowledge. For this reason, the only guide books available for safety cases have assumed expert-level understanding of safety analysis and other relevant topics. Going forward, however, system dependability cannot be achieved without the participation of many ordinary practitioners. Straightforward introductory documents and training courses will therefore be necessary.

- *Development of easy-to-use tools that match the needs of practitioners*: The safety case has only recently started to be put to practical use, and tools such as ASCE from Adelard of the UK remain few in number. The available tools are mainly used to create certification documents, and they cannot easily be integrated with other development tools. As such, they do not adequately reflect the needs of users. Businesses require tools that are easy to use and match their actual needs.

- *Availability of a wide range of notation samples and application case studies*: Given that safety cases contain important corporate information, very few have been made publicly available. Yet, it is difficult to convey to managers and engineers what constitutes an actual safety case. For this reason, access to specific notation samples and case studies is very important.

Figure 4-4 shows the scope of research in the field of the conventional assurance case, and the range of new research topics for D-Case. The assurance case itself is a new area of research where many topics remain unresolved. In terms of both the (conventional) assurance case and the new realm of the

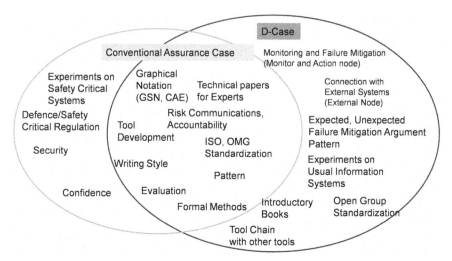

Fig. 4-4 Research Areas for the Conventional Assurance Case and D-Case.

D-Case, this figure shows the types of research we believe to be of particular importance for achieving Open Systems Dependability.

REFERENCES

[1] Bloomfield, R. and P. Bishop. 2010. Safety and Assurance Cases: Past, Present and Possible Future—an Adelard Perspective, Proceedings of the Eighteenth Safety-Critical Systems Symposium, Bristol, UK, 9–11 February 2010, pp. 51–67.

[2] Lord Cullen. 1990. The Public Inquiry into the Piper Alpha Disaster, Vols. 1 and 2 (Report to Parliament by the Secretary of State for Energy by Command of Her Majesty).

[3] Bloomfield, R. and P. Bishop. 1998. A Methodology for Safety Case Development. In Proc. of the 6th Safety-critical Systems Symposium, Birmingham, UK, Feb. 1998.

[4] Leveson, N. 2011. The Use of Safety Cases in Certification and Regulation, ESD Working Paper Series, Boston: MIT.

[5] Eurocontrol, European Organisation for the Safety of Air Navigation. Safety Case Development Manual. European Air Traffic Management, 2006.

[6] Rail Track, Yellow Book 3, Engineering Safety Management, Issue 3, Vols. 1 and 2, 2000.

[7] Ministry of Defense (MoD). 2007. Defense Standard 00-56, Issue 4, Publication Date 01, June 2007.

[8] Kinoshita, Y., M. Takeyama and Y. Matsuno. Dependability Research Report; 2nd February to 9th March; Newcastle, Edinburgh, York, Bath, London UK, AIST Technical Report, http://ocvs.cfv.jp/tr-data/PS2009-002.pdf (in Japanese).

[9] Dependability Research Group, Virginia University. Safety Cases: Repository. http://dependability.cs.virginia.edu/info/Safety_Cases:Repository

[10] The GSN Work Group. 2011. GSN Community Standard Version 1.

[11] Bloomfield, R., P. Bishop, C. Jones and P. Froome. 1998. ASCAD—Adelard Safety Case Development Manual, Adelard.

[12] OMG System Assurance Taskforce, OMG SACM Specification, 2013. http://www.omg.org/spec/SACM/

[13] The Open Group. 2013. O-DA: Open Dependability through Assuredness.

[14] Avizienis, A., J.-C. Laprie, B. Randell and C. Landwehr. 2004. Basic Concepts and Taxonomy of Dependable and Secure Computing, IEEE Transaction on Dependable and Secure Computing, Vol. 1, No. 1.

4.3 D-CASE SYNTAX AND NOTATION METHOD

4.3.1 D-Case Syntax

The standard syntax used for D-Case has been extended from Goal Structuring Notation (GSN)—an assurance-case notation method. Other syntax in, for example, natural language, tabular formats, and formal languages including Agda is defined if providing methods for converting into this standard syntax, and these are to conform to the D-Case syntax.

Figure 4-5 shows a basic example of D-Case description. Here, it is claimed that the system can tolerate response delay failures in a web server, and this is argued on the basis that monitoring for such a failure is possible, and that if a failure is detected, the system can resolve it. D-Script, which will be described in detail later, would be used in practice to resolve such a failure.

Based on the GSN Community Standard, D-Case notation features a number of extensions conceived of during the DEOS development process. Many ambiguities exist within the GSN Community Standard, and certain

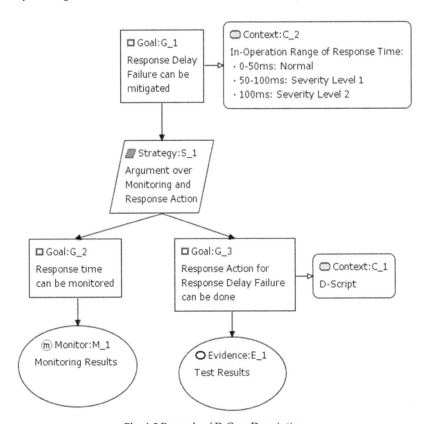

Fig. 4-5 Example of D-Case Description.

definitions remain to be added. The D-Case specification is being formulated based on the GSN Community Standard by the D-Case Committee in conjunction with efforts to resolve ambiguities and complete definitions and also to specify additional features of DEOS. Furthermore, a reference implementation has been realized in D-Case Editor [3] and D-Case Weaver [4].

(1) Node types

D-Case features nodes already defined in GSN as well as its own extension nodes. In addition to descriptive text, a number of different attributes such as responsibility can be defined for each. Figure 4-6 shows the original GSN nodes and the D-Case extension nodes.

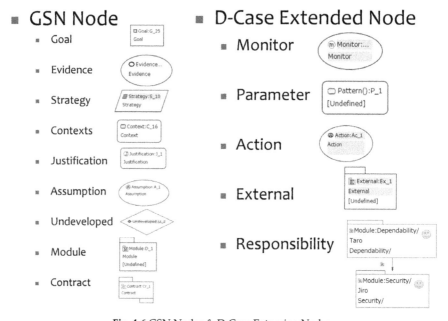

Fig. 4-6 GSN Nodes & D-Case Extension Nodes.

GSN nodes

- *Goal*: A goal node describes a claim to be argued for the system under consideration. Typical examples could be "The system is dependable" and "The system is suitably safe".
- *Evidence*: The purpose of an evidence node is to provide assurance that the claim of its corresponding goal node holds. This could take the form of, for example, a document detailing how a program is implemented, verification results from testing, and/or other formal methods.

- *Strategy*: Strategy nodes describe how the claim in the goal is to be argued to refine it by decomposing it into sub-goals. If, for example, the goal were "The system is safe", we could argue on the basis that the system can mitigate all known threats. Describing the corresponding strategy as "Decompose by identified threat", one of the resulting sub-goals could be "The system can mitigate Threat X".

- *Context*: Context nodes provide information serving as assumptions (or conditions) for the definition of goals and strategies. They might typically describe the operating environment, the scope of the system, or a list of known threats. It is extremely important that the corresponding environment and operating conditions be clarified in this way so that the associated goal or sub-goal can be argued properly.

- *Justification*: The justification node is a sub-class of the context node.

- *Assumption*: The assumption node is a sub-class of the context node.

- *Undeveloped*: An undeveloped node is used whenever the argument or evidence is insufficient to support the goal.

- *Module*: Module nodes provide for the referencing of D-Cases of other modules. This type of node is assigned information, such as the name of the responsible person, as accountability attributes.

- *Contract*: A contract node describes the relationship between two modules. The definition of this node in the GSN Community Standard is particularly ambiguous, and the current specification is under review.

D-Case extension nodes

- *Monitor*: Monitor nodes provide evidence based on the system's runtime data. A typical example of this type of data would be log entries detailing the response speed of a web server. This node is a sub-class of the evidence node.

- *Parameter*: Parameter nodes are used to set parameters within D-Case patterns. This node is a sub-class of the context node.

- *Action*: The role of the action node is to describe an operation procedure (or script) related to system failure response at runtime. This provides the evidence that the procedure will be executed whenever its condition is satisfied. A typical example of this evidence would be countermeasures described in D-Scripts (explained in detail later) for a response delay failure in a web server. This node is a sub-class of the evidence node.

- *External*: External nodes represent modules managed by an external organization. This node is a sub-class of the module node.

- *Responsibility*[1]: Responsibility nodes are used to explain the relationships between modules with different responsibility attributes. They are put to links between modules, for example, as a small node labeled "R" in Fig. 4-6.

(2) *Connecting nodes by links*

As shown in Fig. 4-7, goals are decomposed on the basis of strategies. D-Case leaf nodes can be evidence, undeveloped, module, monitor, action, or external nodes. Context nodes are connected to a goal or strategy.

Two types of links are provided:

- *SupportedBy*: This support link connects from goal to strategy, strategy to goal, goal to evidence, goal to monitor, goal to external, goal to undeveloped, and strategy to undeveloped (Fig. 4-8).
- *InContextOf*: This context link connects from goal to context and strategy to context (Fig. 4-9).

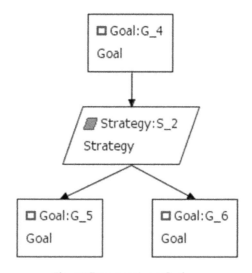

Fig. 4-7 Decomposing a Goal.

---

[1] The difference between accountability and responsibility was the subject of much debate in the DEOS Project. It was ultimately decided that "accountability" applies to the attribute (without boundaries) of dependability of the entire system, while "responsibility" concerns a binomial relationship between sub-systems within the system or a binomial relationship between stakeholders and refers to an obligation (with boundaries) of one with respect to the other. It was argued for the purpose of DEOS that responsibility fulfillment by all system stakeholders is linked to the accountability of the entire system in the case of a closed system, but it is not on its own sufficient for an open system as the boundaries cannot be determined.

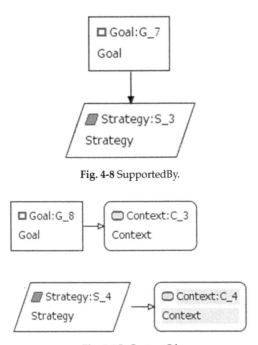

Fig. 4-8 SupportedBy.

Fig. 4-9 InContextOf.

(3) Inter-module relationships

D-Case modules represent individual D-Case descriptions having a single top goal. Inter-module relationships have been defined to show how these modules interact with one another and are represented using double-headed arrows. Figure 4-10 shows an example of inter-module relationships between D-Case modules of a LAN device system.

The d* framework, which will be described in detail in Section 4.5, provides a mechanism by which interdependency relationships between modules can be managed, and this framework has been defined on the basis of the inter-module relationship. For example, parties with responsibility can be assigned on an individual module basis, and responsibility relationships can be allocated between these parties using D-Case inter-module relationships. In the example shown in Fig. 4-11, Taro has responsibility for the *dependability* module; Jiro for the *security* module. The dependability module represents D-Cases related to the dependability of a certain system. The argument supporting dependability also contains an argument that supports security. The *security* module is referenced in order to argue security. A responsibility relationship is required because different parties are responsible for these two modules' results. This relationship is described in the responsibility node labeled "R" in the diagram.

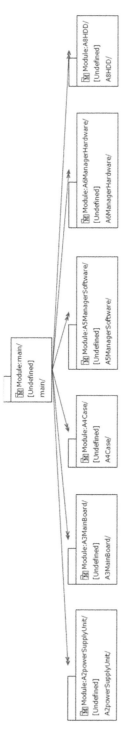

Fig. 4-10 Example of Inter-Module Relationships.

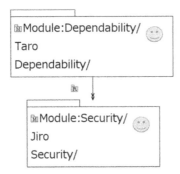

Fig. 4-11 Example of Responsibility Relationship.

4.3.2 D-Case Description Guideline

D-Case descriptions are developed at all stages throughout the life of a system, using many types of document and with the participation of a wide range of stakeholders. Research and development concerning the assurance case is ongoing and highly active, but no standard guideline has yet been provided for widespread application. This section describes the D-Case Description Guideline proposed by Matsuno and Yamamoto [1]. There are nine steps as follows:

1. Clearly define the system's lifecycle and identify the input and output documents for each phase.
2. Categorize the input and output documents.
3. Set "The system is dependable" as the top goal.
4. Attach dependability requirements, environmental information, and lexical definitions to the top goal as contexts.
5. Plan the overall argument structure (D-Case).
6. Attach the necessary documents as contexts and evidence.
7. Develop the D-Case sub-trees based on the documents.
8. Use standard argument structures to add any sub-trees not automatically developed.
9. Repeat the above steps as many times as necessary.

Each of these steps is now described in detail.

Step 1: Clearly define the system's lifecycle and identify the input and output documents for each phase.

D-Case description is to be developed on the basis of documents produced throughout the life of the system under consideration. The reason for this is that D-Case is not intended to replace the documents produced in conventional system development and operation phases; instead, it is fundamentally a document for demonstrating, using those other documents, that the system under consideration is dependable. (Alexander et al. argue that the development of an assurance case does not replace conventional risk analysis and requirement analysis methods [2].) In the future, methods may be developed for first developing the D-Case and then carrying out the actions required to produce the documents throughout the lifecycle of the system.

A simple system lifecycle is shown in Fig. 4-12. Assuming a system with a lifecycle defined as shown in the figure, the input and output documents for the various phases might be as shown in Table 4-1.

Fig. 4-12 System Lifecycle.

Table **4-1** Typical Input & Output Documents over System Lifecycle.

Phase	Input	Output
Requirement definition	User interview transcripts	Requirement definition document
Architecture design	Requirement definition sheets	Architecture specifications, operation definition sheets
Implementation	Architecture specification document	Program code
Test	Program code	Test results
Operation	Operation definition sheets	Operation logs

The system's D-Case would be developed based on these documents. All of the documents produced in developing the system would be included, and a necessary and sufficient set of documents relevant to system dependability must be extracted from these.

Step 2: Categorize the input and output documents.

We must now consider how the extracted input and output documents relate to the system's D-Case—that is, how these documents are used in the argument that supports system dependability. Based on past experience, we would expect the following types of documents to be of relevance to D-Cases:

1. Standards: ISO 26262, ISO/IEC 12207, and other international standards with which the system must comply.
2. Results from risk analysis: Findings of analysis of risks to service continuity—for example, the conclusions of hazard analysis, FTA, and the like.

3. Dependability requirements: A typical example would be "Availability of 99.999%"—dependability requirements vary from system to system and must be defined clearly and individually. In terms of the above example, user interview transcripts and requirement definition sheets would be relevant.

4. Documents concerning the system lifecycle: System lifecycle documents such as described in Step 1 are critical if dependability is to be supported through argumentation.

5. System architecture models: System architectures are described using UML and the like. These must be referenced whenever it is necessary to define how individual system components contribute to dependability. For our example above, the architecture specification document would be of relevance.

6. Operation-related information: Failure occurs while a system is operating; accordingly, details of how the system would be operating at any time constitute critical information. For example, system log data would be important in determining the current dependability status of the information system. Operation definition sheets and operation logs would be relevant in our example.

7. Environment-related information: The dependability of a system can only be supported by argument if the system's environment has been determined.

8. Test and verification results: Information of this nature is referenced using D-Case evidence nodes, and it provides ultimate support for the dependability of the system.

9. Program code: Code developed in order to respond to failure is of particular importance in arguing the dependability of a system.

Step 3: Set "The system is dependable" as the top goal.

Steps 1 and 2 prepare for development of the D-Case, and in Step 3, we are finally ready to get started. The first thing that must be determined is the claim in the top goal. The exact nature of the dependability of a system depends on the system itself and its environment. We should begin, therefore, by setting the generic "The system is dependable" as the top goal. What "dependable" means in this specific context must be defined based on documents concerning the system's dependability requirements. Typical examples might be "The system is sufficiently safe" or "All known threats are being appropriately mitigated in the system"; however, the specific meaning will not be clear at this point in time when work on the D-Case is just starting. Setting a generic claim in a goal allows us to move on despite this restriction. The top goal often becomes clearer as the D-Case is refined.

Step 4: Attach dependability requirements, environmental information, and lexical definitions to the top goal as contexts.

Detailed information on dependability requirements, the system's environment, lexical definitions, and other background required in order to argue the top goal must be attached using context nodes. Ideally, the description of the claim in the goal node should be as simple as possible. When, for example, the top goal is "The system is dependable", referencing the dependability requirement sections of requirement definition documents as context can make the meaning of "dependable" easier to interpret. Information pertaining to the environment of the system under consideration must also be clearly defined, particularly with regard to the scope of the system. If this is not done, the argument will become divergent.

Step 5: Plan the overall argument structure.

Conventional methods have developed dependability cases by setting sub-goals one by one using a deductive approach. Our experience has shown that planning the argument for the entire D-Case early on allows the goal to be decomposed with one eye on the overall structure, and we believe this to be an effective approach. Practical experience with D-Case description has also revealed a number of typical argument structures, some examples of which are as follows:

- Argument based on life cycle
- Argument based on system functionality
- Argument based on system structure
- Argument based on workflow
- Argument based on failure and risk reduction

Planning of the overall argument structure should begin by combining these approaches.

Step 6: Attach the necessary documents as contexts.

Now that the overall structure of the argument has been determined, the documents required for that argument must be attached as contexts and evidence. For example, if an argument structure based on the system's lifecycle has been selected, a context node describing the lifecycle information should be linked to the strategy node (Fig. 4-13).

In another example, a D-Case concerned with failure response is argued. Figure 4-14 shows that the system can tolerate Failure X. The top goal has a definition of Failure X linked as a context node. The argument that Failure X can be mitigated is decomposed on the basis of failure detection and response. The program code for dealing with Failure X is linked as a context node.

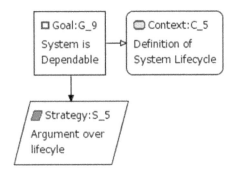

Fig. 4-13 Context Node for Developing an Argument.

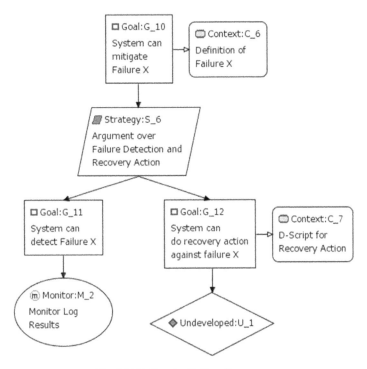

Fig. 4-14 D-Case for Failure Response.

Step 7: Develop the D-Case sub-trees based on the documents.

The argument must now be developed in the context of the documents attached using the context node in Step 6. If this must be done based on the detailed content of the documents, a D-Case sub-tree should be created by developing each of the corresponding documents. The majority of documents can be converted into D-Case in an automatic or semi-automatic fashion. Two examples are presented below, and further details are provided in Section 4.4.

Example 1: Process

Generally speaking, a process is defined as a goal (the objective, the claim), plus steps 1 through N, plus the inputs and outputs for each of the steps. For a process defined in this way, a D-Case sub-tree such as the one in Fig. 4-15 can be automatically created.

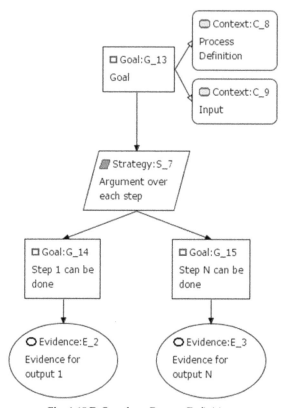

Fig. 4-15 D-Case from Process Definition.

Example 2: Dependability attributes

If dependability is defined from a number of attributes such as availability and reliability, sub-goals can be decomposed based on each of them as shown in Fig. 4-16.

Step 8: Use standard argument structures to add any sub-trees not automatically developed.

After developing sub-trees automatically or semi-automatically by planning the overall structure, attaching the necessary documents as contexts, and decomposing the documents, any sub-trees that were not created must be

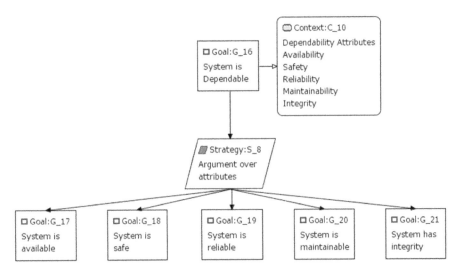

Fig. 4-16 D-Case from Dependability Attribute Definitions.

devised on a personal basis. Our experience has shown, however, that sub-trees created based on an argument structure devised in a totally independent, original manner can be difficult for others to interpret. If development on a personal basis is unavoidable, it is good practice to choose an argument structure that has been successfully applied in the past.

Step 9: Repeat the above steps as many times as necessary.

D-Case comprises informal as opposed to formal arguments, and as such, it is difficult to specify what makes good description. Research into the qualitative and quantitative evaluation of assurance cases is ongoing, and as yet, no particular evaluation method has been widely adopted. This stems from the fact that certain content cannot be assessed in a basic sense according to clearly defined standards. And because the ideal description cannot easily be identified, we should aim to construct arguments in the best way possible so as to develop better D-Cases. One of the most important benefits of D-Case is the way in which describing a system's dependability using D-Case can help to promote better understanding among all stakeholders.

We now apply these steps to a real-world example—a web server system developed for reference purposes at the DEOS Center and shown in Fig. 4-17.

The main components of this system are a web server, an application server, and a database server, which are all controlled by an operator via a console. Users access the system from their client PCs over a network. We now develop the D-Case for this system based on experience gained at the DEOS Center in writing notation.

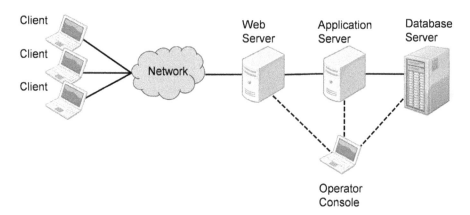

Fig. 4-17 Web Server System.

Step 1: Clearly define the system's lifecycle and identify the input and output documents for each phase.

The web server system in question was developed by integrating existing server PCs. Let us assume that its lifecycle is as shown in Fig. 4-18 and input and output documents are as shown in Table 4-2. It should be noted that because this system was set up for reference purposes only, some of the documents are hypothetical. Here, we focus on the operation workflow definition document in particular.

Fig. 4-18 Lifecycle of the Web Server System.

Table 4-2 Input and Output Documents over the Lifecycle of the Web Server System.

Phase	Input	Output
Requirements definition	User interview transcripts	Requirements definition document, SLA document
Architecture design	Requirements definition document, SLA document	Architecture definition document, operation workflow definition document, risk analysis document
Integration	Architecture design document, operation workflow definition document, risk analysis document, server specifications	Integrated program code
Test	Integrated program code	Test results
Operation	Operation workflow definition document	Operation logs, system logs

Step 2: Categorize the input and output documents.

We now consider how the documents produced over the lifecycle of the system are related to its dependability and categorize them accordingly.

1. *Standards*: As this system was set up for reference purposes, no provisions were made for compliance with international standards and the like. Demonstrating that a system complies with the relevant international standards is one of the principal objectives of the conventional safety case and would be important when developing the D-Case for an actual system.

2. *Results from risk analysis*: In our example, the risk analysis document produced during the architecture definition phase falls under this category.

3. *Dependability requirements*: We can determine the system's dependability requirements based on user interview transcripts, requirement definition documents, service level agreements (SLAs), and other similar material from the requirements definition phase.

4. *Documents concerning the system lifecycle*: This category can consist of the lifecycle definition documents and other relevant input and output documents specified in Step 1.

5. *System architecture models*: The architecture design document falls under this category. It serves as a reference when establishing the scope of the system and arguing according to structure.

6. *Operation-related information*: This category would typically contain the operation workflow definition document, operation logs, and system logs. The latter two play an important role as evidence of how the system is actually operating.

7. *Environment-related information*: As this particular system was set up for reference purposes, no consideration was given to actual environment-related information. When working with an actual system, however, this information must be combined with the above operation-related information in order to develop an argument concerning how, and in what specific conditions, the system is operating.

8. *Test and verification results*: Test results from the test phase fall under this category.

9. *Program code*: In our example, this category would contain the integrated program code.

Step 3: Set "The system is dependable" as the top goal.

While the service under consideration is intended for end users, we assumed that the order for development would be placed by the web service provider. A D-Case should be produced from the perspective of the end user, but for the purpose of this example, we assumed that the system developer would use the D-Case to assure the web service provider that the developed system is dependable. When working with an actual system, consideration would also need to be given to other stakeholders such as the web server operator. A web server developer and a web service provider would normally agree upon non-functional requirements in the form of an SLA. For this reason, we redefine the top goal as "The web server complies fully with the SLA".

Step 4: Attach dependability requirements, environmental information, and lexical definitions to the top goal as contexts.

We now link context nodes to identify information that provides context for arguing the top goal—that is, "The web server complies fully with the SLA". Specifically, we attach the architecture design document required for the argument; the fact that the system scope comprises a web server, an application server, and a database server; and the SLA document to provide the details of the agreement. Having completed all of the steps above, our D-Case now looks like that in Fig. 4-19.

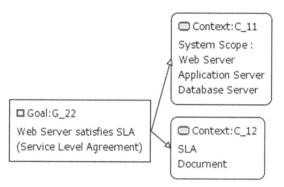

Fig. 4-19 Top Goal with Context Nodes.

Step 5: Plan the overall argument structure.

For our example, we plan the D-Case structure in line with the following approach to the overall argument:

> *The server PCs were purchased and integrated instead of developing them ourselves; accordingly, the detailed technical nature of their internal components cannot easily be confirmed. Furthermore, given that modern PC technology is considerably advanced, failure due to the malfunction of internal components should not constitute a major cause for concern—particularly in regard to small-scale systems such as this. Accidents caused by human error when operating servers and resulting in large-scale loss of information have, however, been seen in recent times. We must, therefore, develop an argument to address all conceivable failures as identified through risk analysis in each step in the overall operation workflow. Having done this, we develop arguments for rapid failure response and change accommodation.*

This approach to arguing according to rapid failure response and change accommodation is the product of much debate during the course of the DEOS Project, and it is based on the belief that the system must respond as quickly as possible in the event of a failure and that any change in the system itself or in its environment must be accommodated. Figure 4-20 illustrates the overall argument structure in D-Case arrived at in this way.

Fig. 4-20 Overall Argument Structure for Web Server System.

Step 6: Attach the necessary documents as contexts.

In order to develop an argument in line with the operation workflow, the operation workflow definition document must be attached using a context node. From this document, we determine that the operation workflow comprises a total of six steps—user login, shopping cart processing, credit card authentication, finalizing, delivery, and customer service. In addition, the document defines these steps as being further refined into sub-steps.

Step 7: Develop the D-Case sub-trees based on the documents.

In our example, we can decompose the D-Case top goal for each step in accordance with the operation workflow definition document as shown in Fig. 4-21 (only the sub-trees for the first and last steps are shown).

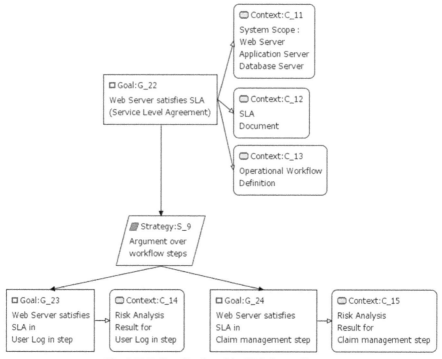

Fig. 4-21 D-Case Top Level for Web Server System.

REFERENCES

[1]   Matsuno, Y. and S. Yamamoto. 2013. A New Method for Writing Assurance Cases, IJSSE 4(1), pp. 31–49.
[2]   Alexander, R., R. Hawkins and T. Kelly. 2011. Security Assurance Cases: Motivation and the State of the Art, Technical Note CESG/TR/2011/1, High Integrity Systems Engineering, Department of Computer Science, University of York.
[3]   http://www.dcase.jp/editor_en.html
[4]   http://www.jst.go.jp/crest/crest-os/tech/DCaseWeaver/index-e.html

4.4 ROLES PLAYED BY D-CASE

D-Case can play the following five roles in consensus building and achieving accountability, both of which are critical elements for assuring the dependability of a system:

- Explicitly defining the claim that must be argued.
- Explicitly defining evidence that serves as the basis for supporting the claim.
- Explicitly defining the context for the argument.

- Logically arguing the claim using the evidence.
- Supporting consensus building in an objective manner through the use of standardized notation.

To demonstrate these roles, the following describes how a D-Case D would be used to achieve accountability for a claim C concerning a target system T. C is the top goal of the D-Case D. We assume the correspondence relationship between D and T is consistent; if not, the D-Case would need to be corrected.

We can make a recursive argument along the length of the relationship (the depth of the tree) from $C0$—the top goal position—to the evidence. We use the accountability proof sequence A (presented below) to argue systematically the current claim CC, positioning it as the top goal $C0$ for system T. If the result of A is agreed upon, then accountability is achieved. If there is disagreement—which would indicate a problem with the D-Case evidence or comprehensiveness of its argument—then accountability is not achieved. Failure to achieve accountability would be resolved by either (1) confirming that system T is free of problems and redeveloping the D-Case, or (2) identifying and correcting any problems in system T and redeveloping the D-Case.

Accountability Proof Sequence A

Input   : The current claim CC, and a D-Case D with CC as the top goal

Output result : Agreement or disagreement

Process   : As follows:

The subordinate nodes connected to the current claim CC are either evidence or strategy nodes.

(1) When the node subordinate to CC is evidence E:
     If CC is directly linked to evidence E, the claim can clearly be supported according to the validity of E. Once agreement is reached on evidence E, the argument sequence A for CC is ended, with agreement as the result. If that is not the case, then argument sequence A is ended with disagreement as the result.

(2) When the node subordinate to CC is strategy S:
     If CC is linked via strategy S to a number of subordinate goal nodes ($SC1$, $SC2$...SCk), the following recursive argument is made.
     (2-1) The completeness of the decomposition to subordinate goals $SC1$ through SCk according to strategy S is argued based on the context node linked to the strategy. If there is a problem with completeness, then accountability proof sequence A is ended with disagreement as the result. If not, the procedure in (2-2) is carried out.
     (2-2) CC is argued for all of the subordinate goals $SC1$ through SCk. First, let $j := 1$.

(2-2a) Repeat the following.
  If $j > k$, then the argument for all of the subordinate goals is complete, so accountability proof sequence A is ended with agreement as the result.
  If $j < k + 1$, then
  Accountability proof sequence A is carried out with SCj as CC.
If the result is agreement, then the procedure in (2-2a) is carried out with $j := j + 1$.
If the result is disagreement, then accountability proof sequence A is ended with disagreement as the result.

End of Accountability Proof Sequence A

We now explain accountability proof sequence A in terms of the example in Fig. 4-22. We start by setting $G1$ as CC. The node subordinate to $G1$ is the strategy node$S1$, so the argument sequence is repeated for the subordinate nodes to $S1$ ($G2$, $G3$, $G4$). Here we determine whether the decomposition of $G1$ to $G2$, $G3$, and $G4$ is complete, based on whether or not it is supported by the context node $C1$ linked to $S1$. If the decomposition to $G2$, $G3$, and $G4$ is not complete, no further proof is valid. If that decomposition is complete, the proof sequence is repeated recursively.

  We look at $G2$ first. Here the node subordinate to $G2$ is the strategy node $S2$, so we determine whether the decomposition to the nodes subordinate to $S2$ ($G5$, $G6$) is complete, based on the context node $C2$ linked to $S2$. If the completeness of decomposition to $G5$ and $G6$ cannot be substantiated by $C2$, then no further proof is valid. If the decomposition to $G5$ and $G6$ is complete, the proof sequence is repeated recursively.

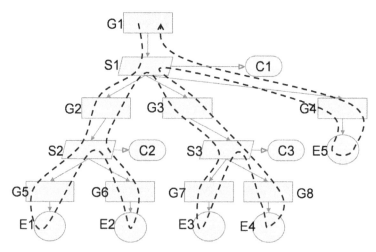

Fig. 4-22 Example for Accountability Proof Sequence.

We now turn our attention to *G5*. Here, the node subordinate to *G5* is the evidence node *E1*, and if it is agreed that *G5* is supported by *E1*, then the proof sequence for *G5* is complete. Likewise for *G6*, the question is whether this goal can be supported by evidence node *E2*. If there is agreement, the proof sequence reverts to the superior node. Now that agreement has been reached for all subordinate goals of *G2*, the procedure again reverts to the superior node and restarts from *G3*. As in the case of *G3*, if it is agreed based on the context node *C3* that decomposition to *G7* and *G8* is complete and that these subordinate goals are supported by the evidence nodes *E3* and *E4*, then agreement is deemed to have been reached for *G3*. Finally, if it can be agreed that *G4* is supported by the evidence node *E5*, then all goals subordinate to *G1* (*G2*, *G3*, and *G4*) will have been agreed upon and the proof is deemed complete.

As we have seen above, the completeness of decomposition of a goal into subordinate goals and evidence must be carefully argued. We proposed six standard decomposition patterns based on how a target system is argued, and these are shown in Table 4-3 in Section 4.6.

In the above explanation, no consideration is given to a goal that cannot be refined in the accountability proof sequence *A*. This is because agreement is not possible when a goal cannot be decomposed into sub-goals with their validity supported by evidence. Note that there may be cases where agreement is sought regarding non-refinement. With respect to assuring dependability, however, that would constitute withholding of a decision, so it is not considered to be a dependability-related agreement. In other words, it would put agreement concerning dependability on hold, and in terms of accountability proof sequence *A*, would constitute a decision not to agree on the D-Case.

4.5 d* FRAMEWORK

Management of system interdependence is highly important in recent systems where externally developed components have been employed and/or services provided by external systems are utilized [1]. In order to assure the dependability of a given system *A*, it would be necessary to assure not only the dependability of system *A* itself, but also that of external systems with which system *A* has interdependence relationship and that of the interdependencies with these systems.

We now describe how interdependencies such as this can be managed using D-Case. For the purpose of explanation, we assume that system *A* and system *B* mutually interact, and system *C* is a sub-system within the ownership of system *A*. The D-Case for system *A* is denoted by $d(A)$, which assures the dependability of system *A* itself. Next, we define $d(A,B)$ as the D-Case necessary for demonstrating that system *B* satisfies the dependability requirements of system *A*, with which it interacts. This type of D-Case is referred to as an inter-dependability case. Similarly, we define $d(A,C)$ as the D-Case necessary for demonstrating that system C satisfies the dependability requirements of system A, with which it interacts. We also consider d(B) and d(C) as the D-Cases that must be developed to assure the dependability of system B and system C themselves, respectively.

These interdependencies are shown in Fig. 4-23. Because the interdependency relation shows the transitive property, this diagram is referred to as a *d* framework* [2]. Here, each one of the D-Cases is shown as a tree, the root of which represents the corresponding dependability goal. From this d* framework, we can see that the dependability of system A is assured by combining d(A), d(A,B), d(A,C), d(B), and d(C). The dotted line encloses the range of D-Cases showing the ownership of system A. A D-Case node can be replaced with a D-Case tree having an interdependence relationship, and this allows the propagation of dependability between interdependent systems to be represented. In cases where components are procured, a D-Case for each procured component together with one for the interdependency between the procured components and the system configured from the components must be developed. Component suppliers should develop component D-Cases to assure the required level of internal dependability. Meanwhile, the party that procures these components must develop D-Cases that assure dependable usage of the components in order to assure the dependability of the integrated system. It will be necessary to integrate the internal D-Cases and D-Cases for interdependencies with one another in order to verify consistency between them. Using a d* framework, the provider of the integrated system can assure the dependability of that system and can also confirm whether components are being efficiently procured.

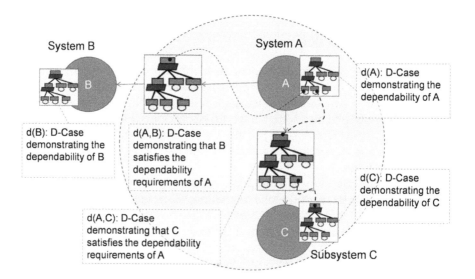

Fig. 4-23 Example of d* Framework.

It is possible that D-Cases will not be available for components procured for external sources—that is, black-box components. For this reason, these D-Cases must be developed in order to assure dependability. The items to be verified for this purpose must be determined based on the usage conditions of the corresponding black-box components.

When investigating inter-system dependability using D-Cases as described above, the following fundamental issues must be resolved in regard to relationships with entities (or actors) bearing system-related responsibility:

  i. How to handle inter-actor responsibilities;

 ii. How to handle consistency between inter-actor responsibilities and dependability; and

iii. How to achieve accountability.

The term "actor" as used here could mean a person or an organization, or alternatively, a system, sub-system, component, or the like. Accordingly, a relationship between actors could take the form of, for example, the relationship between a system and a sub-system or the relationship between a company that orders a system and the one that develops it in response.

A d* framework can be used to describe the dependability relationships between these actors [5]. Such a d* framework contains the following two types of D-Case:

  1. D-Cases assuring the dependability of the actors themselves; and

  2. D-Cases demonstrating that one actor discharges its responsibilities with respect to another actor.

Consider, for example, a set of sensors for monitoring a LAN device together with a LAN device management system that receives LAN device data from the sensors and displays it to operators in order to shut off network access from inappropriate LAN devices. This particular system comprises a LAN device management system, LAN device monitoring sensors, and an operation sub-system. Figure 4-24 illustrates the corresponding d* framework. Here, the three component elements of the system are described as D-Case modules, and each relationship between two of these modules are described as a D-Case goal. (Relationships $d(A,B)$ and $d(A,C)$ in Fig. 4-23 are described as Responsibility nodes put to corresponding inter-module links such as shown in Fig. 4-11, or can interchangeably be described as goals in between modules as shown in Fig. 4-24.)

A D-Case can be developed for each module, and Fig. 4-25 shows the result of mapping the subordinate D-Cases to the modules. Using a d* framework in this way makes it possible to evaluate the dependability of the overall system.

We now look at how a d* framework can be used to confirm inter-organization dependability. A service-related organization, for example, has an inter-organization relationship structure and a relationship structure for

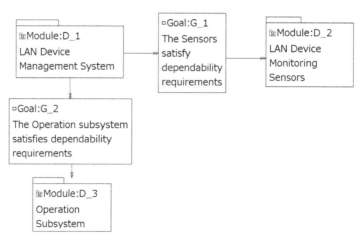

Fig. 4-24 d* Network for LAN Device Management System.

assuring service dependability in a cooperative manner. Figure 4-26 shows a possible organization structure for development of safe services, with organizations identified by circles. Here, customers use the service provider's services. The service provider develops services using the component developer's components. A third-party organization evaluates the safety of the components, as well as the safety of the services. The arrows connecting organizations indicate the existence of these relationships. Specifically, the organization at the tip of the arrow has a responsibility to the organization at the base—for example, the service provider has a responsibility to the customers to provide services safely. The rectangles on the arrows indicate the objective that must be achieved in accordance with each responsibility.

We now show how organization structures and the corresponding responsibility relationships can be displayed using a d* framework. In Fig. 4-27, which transposes the example from Fig. 4-26, each organization is mapped to a module, and the responsibility relationships between the organizations are mapped to goals between the modules. Each of the goals from Fig. 4-27 can be decomposed with subordinate sub-goals. In addition, the inter-organization responsibility relationships are explicitly defined using responsibility attribute links in the form of double-headed arrows.

By displaying responsibility-bearing actors using D-Case modules in this way, we can use goals to illustrate the responsibilities they must discharge. This module-based approach to defining actors with responsibilities not only allows us to develop the relevant arguments and organize evidence on an actor-specific basis, but also makes it easier to understand the relationships between actors bearing responsibility.

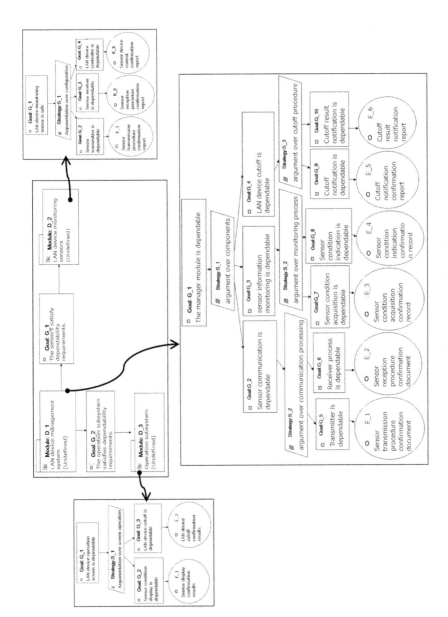

Fig. 4-25 d* Framework Showing Subordinate D-Cases.

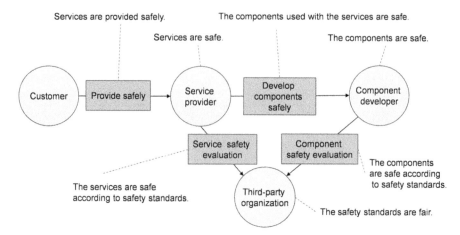

Fig. 4-26 Typical Organization Structure.

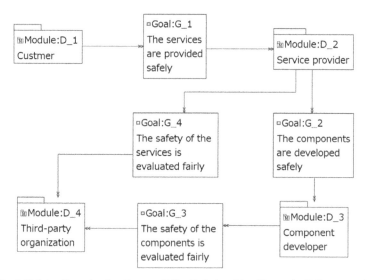

Fig. 4-27 Inter-Organization Responsibility Relationships Shown in d* Framework.

REFERENCES

[1]   Tokoro, M. (ed.). 2013. Open Systems Dependability—Dependability Engineering for Ever-Changing Systems, CRC Press.
[2]   Yamamoto, S. and Y. Matsuno. 2012. d* framework: Inter-Dependency Model for Dependability, IEEE/IFIP Int'l Conf. on dependable Systems and Networks (DSN 2012).

4.6 D-CASE PATTERNS

In order to develop a D-Case, we must first list the dependability-related goals that should be supported for the system. To this end, strategy nodes are used to decompose goals into subordinate sub-goals. At this point, the following questions often arise:

1. What should the goal be and how should it be described?
2. What strategies should be described?
3. How much should the argument be decomposed in line with strategies?
4. What contexts should be described?
5. What evidence should be described?
6. How far should the tree structure be developed?
7. How should the relationships between contexts and evidence be analyzed?

One effective means of resolving these questions is to limit the scope of application to a specific field, allowing us to define in advance how the hierarchical structure and constituent elements of D-Cases for that field should be described [1, 2].

However, in cases where the scope cannot be limited in this way, a more general method is needed. For example, we could conceivably develop D-Cases based on development documents, operation and maintenance manuals, and other existing material. One advantage of this type of approach is that the makeup and contents of the selected documents help to clarify how the D-Case structure and constituent elements should be described.

In the absence of any clearly delineated approaches for decomposing goals in line with strategies and decomposition sequences, systems owners have sometimes not known how to develop their arguments. In response, Bloomfield et al. introduced seven decomposition patterns for safety cases—architecture, functional, set of attributes, infinite set, complete, monotonic, and concretion [3]. However, it is not clear at present whether this list is exhaustive and includes all possible patterns.

When developing a D-Case, we need (a) a target for dependability assurance, (b) an argument to be substantiated, and (c) supporting evidence. In addition, we can also reuse (d) D-Cases whose argument has already been agreed upon.

Hence D-Case patterns may include patterns for the dependability assurance target, patterns for the argument to be substantiated, patterns for evidence, and patterns for how existing D-Cases should be reused. Assurance target patterns can take the form of reference-model decomposition patterns based on the shared architecture of the target, or target-description decomposition patterns based on the way in which the target is described. Argument patterns, meanwhile, can take the form of condition decomposition patterns based on argument conditions, or deductive decomposition patterns based on deductive methods. The relationships between these patterns are shown in Fig. 4-28.

When arguing that a certain system is dependable using a D-Case, it is important that (a) the assurance target and the D-Case are consistent and (b) the D-Case argumentation method is valid.

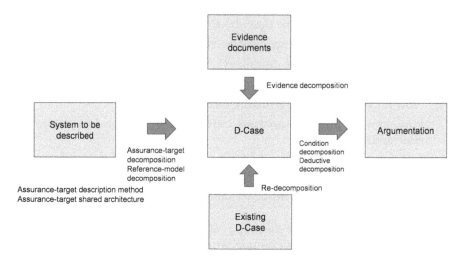

Fig. 4-28 Relationships between D-Case Patterns.

(1) Consistency of assurance target and D-Case

The dependability assurance target and the D-Case developed to argue the dependability of the target must be consistent with one another. Otherwise, the assurance argument may not be applicable even if the developed D-Case is valid, and with such a non-alignment of the arguments, accountability could not be achieved for a goal that the assurance target is dependable.

The assurance target whose dependability is to be assured using the D-Case could be (a) a descriptive structure representing a system or (b) a reference model on which a system is based. In case (a), the description structure representing the system to be assured and the D-Case would be consistent with one another. In case (b), the reference model for the system to be assured and the D-Case would be consistent. A goal argued with such a D-Case could thus achieve accountability in either way.

(2) Validity of the D-Case argumentation method

To argue a goal using a D-Case, it must be decomposed into sub-goals, which requires that we select (a) a deductive method, (b) evaluation conditions, and (c) the type of evidence to support the goals. In addition, (d) the reuse of existing D-Cases with which accountability has been achieved can also make it possible to achieve accountability for the new goal. Let's take a look at why each of these is required.

(a) An appropriate deductive method must be selected because if a unique method were to be used in the D-Case, we would first have to establish its validity. Accountability cannot be achieved if the D-Case uses a deductive method that has no validity.

(b) Suitable evaluation conditions must be selected because decision-making on the basis of conditions concerning the goal must be shown to be valid. Accountability cannot be achieved if the D-Case uses decision-making that has no validity.

(c) Evidence supporting the goal must be carefully selected in order to demonstrate that the selection is valid. A goal cannot be substantiated by evidence that is not appropriate for that specific purpose.

(d) The reuse of existing D-Cases with verified validity eliminates the need to redo certain parts of the overall argument and can reduce rework.

Valid arguments can, therefore, be made in a logical fashion with a D-Case developed using (a) a deductive method, (b) evaluation conditions, (c) effective evidence, and (d) existing D-Cases that each have established validity.

To make it easier to select an optimal structure for decomposition of the D-Case top goal into sub-goals and evidence, Table 4-3 presents six typical patterns for achieving accountability [1, 2]. Different patterns have been prepared for various types of assurance target. For example, if system dependability must be assured from the perspective of system configuration, the Architecture Decomposition pattern can be used.

Table 4-3 Types of Patterns for D-Case Goal Decomposition.

Type of pattern	Basis for goal decomposition
Target Description	Assurance target's description method.
Reference Model	Assurance target's reference model.
Conditional	Conditions relating to the assurance target
Deductive	Argument by *reductio ad absurdum* or inductive reasoning
Evidence	Evidence
Re-Decomposition	Reusing related arguments

Target Description decomposition patterns are used to decompose a goal based on the assurance target's representation structure. Specific patterns from each of these six types are presented in Table 4-4 through Table 4-9. Here, the sixth and seventh examples for the Condition type—namely, Improvement Decomposition and Refinement Decomposition—correspond respectively to the monotonic and concretion patterns proposed by Bloomfield et al. [3].

Table 4-4 Examples of Target Description Decomposition.

	Decomposition pattern	Basis for goal decomposition
1	Architecture Decomposition	System configuration
2	Functional Decomposition	System functionality
3	Attribute Decomposition	Decomposing a property into several attributes
4	Complete Decomposition	All elements of the assurance target
5	Process Decomposition	Process input, action, and output
6	Process Relationship Decomposition	Preceding-process and subsequent-process relationships
7	Level Decomposition	Target level structure
8	DFD Level Decomposition	DFD level structure
9	View Decomposition	UML view structure
10	Use Case Decomposition	Use cases
11	Requirement Description Decomposition	Requirement description items
12	State Transition Decomposition	State transitions
13	Operation Requirement Description Decomposition	Operation requirement definition forms
14	Sequence Decomposition	Sequence diagrams
15	Business Process Decomposition	Business process model notation

Table 4-5 Examples of Reference Model Decomposition.

	Decomposition pattern	Basis for goal decomposition
1	Risk Response Decomposition	A system risk reference model
2	Embedded Reference Model Decomposition	An embedded system reference model
3	Common Criteria Decomposition	Common criteria (CC) for security
4	Requirement Specification Description Decomposition	Chapter structure of requirements-related documents
5	System Environment Decomposition	The information system's environment
6	Failure Mode Analysis Decomposition	Failure mode analysis of the target
7	Non-Functional Requirement Index Decomposition	Non-functional requirement quality index
8	DEOS Process Decomposition	The DEOS Process
9	Test Item Decomposition	Test-item reference model
10	Problem Frame Decomposition	A problem frame pattern

Table 4-6 Examples of Conditional Decomposition.

	Decomposition pattern	Basis for goal decomposition
1	ECA Decomposition	Event, condition, action
2	Conditional Judgment Decomposition	Conditional judgment
3	Alternative Selection Decomposition	Alternative selection
4	Conflict Resolution Decomposition	Conflicts and their solutions
5	Balance Decomposition	Mutually dependent attributes
6	Improvement Decomposition	Improvements from an older system in the new system
7	Refinement Decomposition	Removal of vagueness

Table 4-7 Examples of Deductive Decomposition.

	Decomposition pattern	Basis for goal decomposition
1	Induction Decomposition	Partitioning of the assurance target
2	Elimination Decomposition	Use of the elimination method
3	Negative Reasoning Decomposition	An approach to negation of the goal
4	Rebuttal Decomposition	Evidence refuting the goal

Table 4-8 Examples of Evidence Decomposition.

	Decomposition pattern	Basis for goal support
1	Review Decomposition	Review findings
2	Evaluation Decomposition	Checklists and voting results
3	Test Decomposition	Test findings
4	Certification Decomposition	Formal certification
5	Model Checking Decomposition	Model check results
6	Simulation Decomposition	Simulation findings
7	Agreement Decomposition	Agreement-related documents
8	Monitoring Decomposition	Monitoring results
9	Documentation Decomposition	Documents related to development and operation
10	Legal System Decomposition	Documents related to standards, the legal system, and so forth

Table 4-9 Examples of Re-decomposition.

	Decomposition pattern	Description
1	Horizontal Decomposition	Decomposition, using multiple arguments developed independently from one another, of a number of sub-goals that have been decomposed subordinate to a common goal
2	Vertical Decomposition	Decomposition of a single goal using developed arguments

As an example of how decomposition patterns can be applied, we look at the procedure for developing a D-Case for assuring the dependability of the LAN device monitoring system shown in Fig. 4-29.

First, we develop the subordinate D-Case shown in Fig. 4-30 using the DEOS Process decomposition pattern, which was used to describe the *DEOS Process* in D-Case in Section 3.4. In this figure, LDMS stands for "LAN device monitoring system".

Next, we apply the Architecture Decomposition pattern to the goal "Ordinary Operation of the LDMS is achieved", giving the D-Case shown in Fig. 4-31. We can see from Fig. 4-29 that the LAN device monitoring system comprises sensor management and sensor sub-systems; accordingly, the dependability of the system can be argued in terms of the dependability of these two sub-systems and also the dependability of the interactions between them.

In this way, D-Cases can be efficiently developed by applying decomposition patterns in a recursive fashion [4].

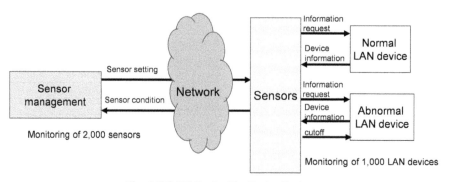

Fig. 4-29 LAN Device Monitoring System.

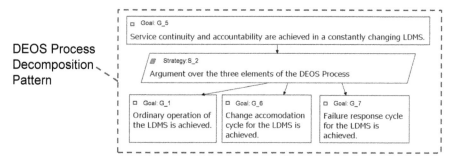

Fig. 4-30 D-Case for LAN Device Monitoring System.

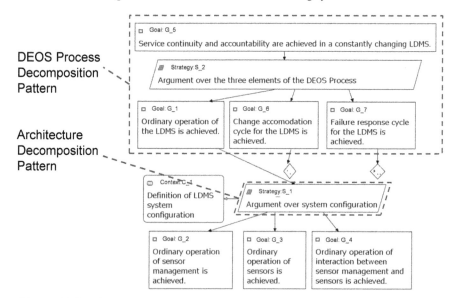

Fig. 4-31 D-Case for LAN Device Monitoring System based on Architecture Decomposition.

REFERENCES

[1] Matsuno, Y. and S. Yamamoto. Practical D-Cases (in Japanese).
      (http://ec.daitec.co.jp/finditem.aspx?scode=978-4-862930-91-0)
[2] Yamamoto, S. Claims & Evidence (in Japanese).
      (http://ec.daitec.co.jp/finditem.aspx?scode=978-4-86293-095-8)
[3] Bloomfield, R. and P. Bishop. 2010. Safety and assurance cases: Past, present and possible
      future—an Adelard perspective. Proceedings of 18th Safety-Critical Systems Symposium,
      February 2010.
[4] Yamamoto, S. and Y. Matsuno. 2013. An Evaluation of Argument Patterns to Reduce
      Pitfalls of Applying Assurance Case, 1st International Workshop on Assurance Cases for
      Software-intensive Systems (Assure 2013).

D-Case Tools

We have developed various tools to facilitate the use of D-Case. In this chapter, D-Case Editor and D-Case Weaver are described, followed by the current research status and challenges being undertaken to promote wider use of D-Case and its tools.

5.1 D-CASE EDITOR

To encourage the uptake of D-Cases and also to promote practical implementation thereof, we have developed D-Case Editor—a D-Case tool that is fully integrated with other DEOS solutions. D-Case Editor has been made publicly available as freeware. Currently, it provides the following functionality:

- Editing of graphical D-Case notation
- Consolidation of D-Case pattern libraries, which are collections of reusable D-Case descriptions
- Conversion to OMG ARM and SACM metamodels
- Real-time monitoring of a target system using monitor and action nodes (in coordination with D-RE)
- Linking with a benchmark environment for dependability measurement (in coordination with DS-Bench [1])
- Integration with Agda verification functionality (see Chapter 6)

D-Case Editor takes the form of a plug-in for the Eclipse[1] open-source integrated development environment. Accordingly, it can be easily extended using the Eclipse plug-in development framework. Figure 5-1 shows a typical D-Case Editor screen.

---

[1]http://www.eclipse.org/

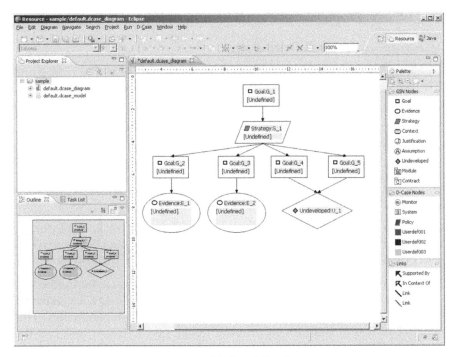

Fig. 5-1 Typical D-Case Editor Screen.

(1) Management of D-Case-related Documents

In the practical use of D-Cases, we need many related documents that are linked with them. Thus, management of D-Case-related documents is equally important. Documents constituting evidence, contexts, and the like must be easy to reference from a D-Case so that the dependability argument can be readily understood and the stakeholders can reach a consensus. Many different stakeholders contribute to the production of these documents, and they are updated frequently over the lifecycle of the system; accordingly, these D-Case-related documents need to be properly managed. They must be registered in a document repository that can be accessed by a range of stakeholders; meanwhile, other administrative functions such as version management are also required.

Thus, D-Case Editor has been extended to allow the linking of D-Cases with documents registered in a document repository and subjected to version management. We refer to this functionality as *D-Case document control*. The following describes the workflow associated with the D-Case document control function and how it can be put to use in practice.

(2) D-Case Document Control Workflow

D-Cases are used in the following steps:

- Developing a dependability argument using the D-Case approach and clarifying both the goal to be supported and the scope of the argument.
- Preparing graphical D-Case description in order to present the argument and the corresponding reasoning in graphical format.
- Supporting the goal with evidence documents to provide assurance.
- Achieving accountability, when necessary, using the graphical D-Case description and the related evidence documents.

There are five actors who carry out these D-Case activities:

Operations Department: The department and personnel responsible for the development of products or the provision of services carry out D-Case activities in accordance with the development processes applied in their organization.

D-Case Manager: The scope of each D-Case and its context are defined by a manager, who also produces the context documents. The D-Case manager ensures that accountability can be achieved in terms of dependability. This manager is responsible for approving developed D-Cases and has the Operations Department produce evidence documents.

D-Case Developers: D-Case Developers interpret work processes implemented according to their organization's methodology and develop D-Cases from the perspective of dependability, with scope and context taken into consideration.

Document Management System: This system houses scope documents, context documents, and operations results as evidence documents, as well as the D-Cases to which they are linked. It can present for review and otherwise disclose these documents to stakeholders. The management system also keeps track of the approval status of D-Cases.

D-Case Repository

The D-Case Repository is used to register D-Cases under development and to implement structure management.

D-Case document control flow consists of the following steps:

(1) D-Case development and correction

D-Cases are developed for the product or service under development. In this step, the scope of each D-Case and its context are defined, context documents are linked to context nodes, and a prototype D-Case description is created.

(2) Linking evidence documents to the D-Case

The documents prepared by the Operations Department and housed in the Document Management System are linked to evidence nodes. In this step, the evidence documents that must be linked to evidence nodes are identified and the task of preparing and storing the evidence documents is assigned to the Operations Department.

(3) Reviewing related documents and the D-Case

Once documents have been linked to context and evidence nodes, the D-Case is reviewed. In this step, evidence documents are checked to ensure that they are the latest versions, the D-Case is registered in the Document Management System, and the related documents are referenced from the D-Case in order to ensure that the evidence is valid. In the case of re-review, different versions of the D-Case are compared in order to confirm whether all the changes have been properly reflected.

(4) Approving the D-Case

The reviewed D-Case and related documents are approved within the Document Management System. In this step, the D-Case is approved when review is complete.

Depending on the processes employed by the development organization in question, these steps may be implemented either in a waterfall-type pattern or repeated in an iterative fashion.

The above steps are now described in detail with the help of workflow diagrams for each. Here, the actors are arranged vertically, and tasks are executed from left to right. Figure 5-2 shows the meaning of the symbols used in the diagrams.

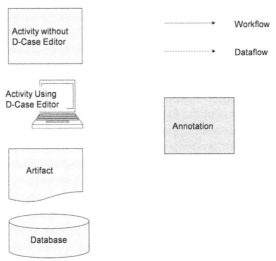

Fig. 5-2 Symbols Used in D-Case Workflow Diagrams.

Figure 5-3 illustrates **Step 1:** *D-Case document development workflow*. First of all, the dependability requirements need to be agreed by the Operations Department, the D-Case Manager, and the D-Case Developers. The D-Case Manager then prepares context documents on the basis of the agreed requirements and registers them in the Document Management System. These context documents typically contain service objective definitions, stakeholder definitions, requirement items, service continuity scenarios, system requirements, and so forth. Next, the D-Case Developers refer to the context documents in order to create graphical D-Case description. As part of this process, context nodes are added to the D-Case in order that context documents from the Document Management System may be referenced, thereby clarifying the overall D-Case goal and scope. The D-Case is then developed through to leaf sub-goals in order to produce a prototype for the argument. All work done during this step is then reviewed internally within the development organization, and the D-Case is stored in the D-Case Repository.

Based on the prototype produced as described above, **Step 2:** *Linking evidence documents to the D-Case* can now be carried out as shown in Fig. 5-4. The Operations Department, the D-Case Manager, and the D-Case Developers come together to make arguments on the basis of the D-Case registered in the D-Case Repository and thereby identify documents constituting evidence. Based on their agreements, the Operations Department prepares the actual evidence documents, which can take the form of design specifications, test specifications, test results, review reports, and the like. When all of these documents have been completed, they are registered in the Document Management System. The D-Case Developers then examine these documents and link them to the D-Case as evidence.

When evidence documents have been prepared and registered as described above, Step 3: *Reviewing related documents and the D-Case* can get under way as shown in Fig. 5-5. In this step, the D-Case Developers determine whether the evidence documents must be updated, and if necessary, they perform the necessary D-Case update work to ensure overall consistency. When this work comes to an end, the D-Case is registered in the Document Management System, and the Operations Department and D-Case Manager jointly review the related documents and the D-Case.

When the review is over, approved status is assigned to the D-Case in the Document Management System (Fig. 5-6). This means that, using the D-Case, the development organization is ready for achieving accountability to outside stakeholders.

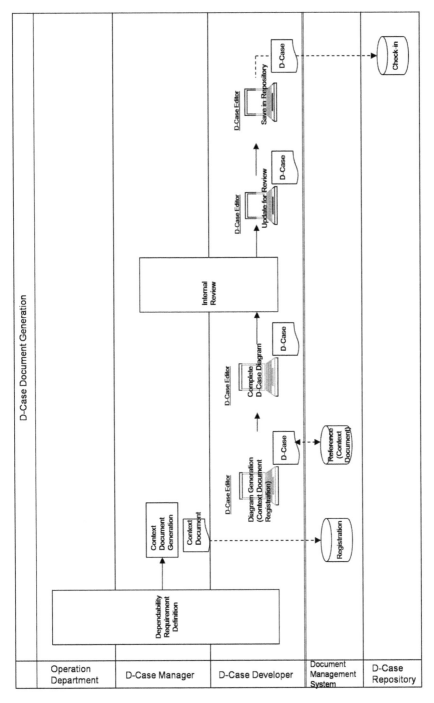

Fig. 5-3 D-Case Document Generation Workflow.

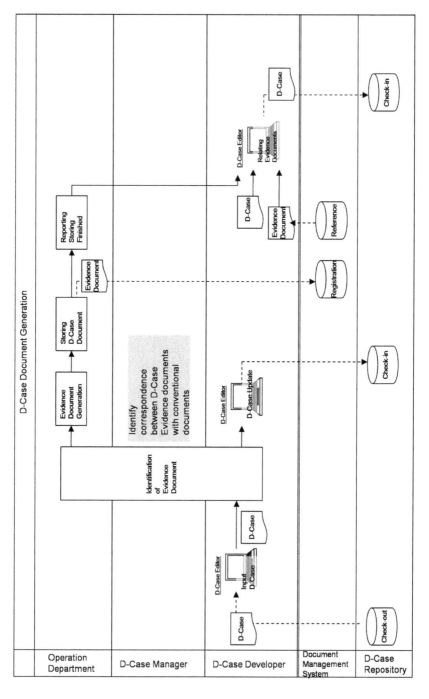

Fig. 5-4 Linking Evidence Documents to D-Case.

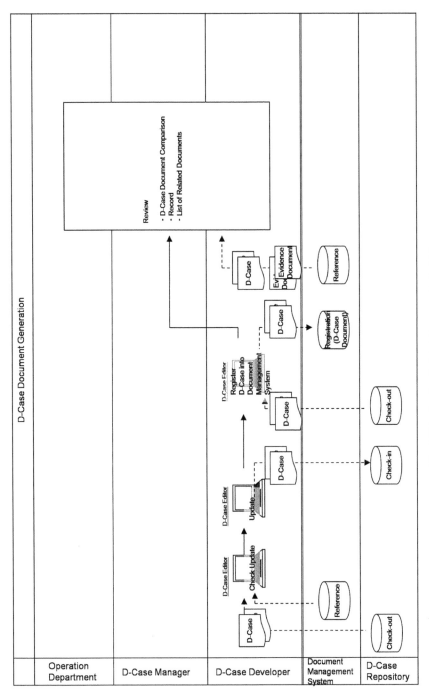

Fig. 5-5 Reviewing Related Documents & D-Case.

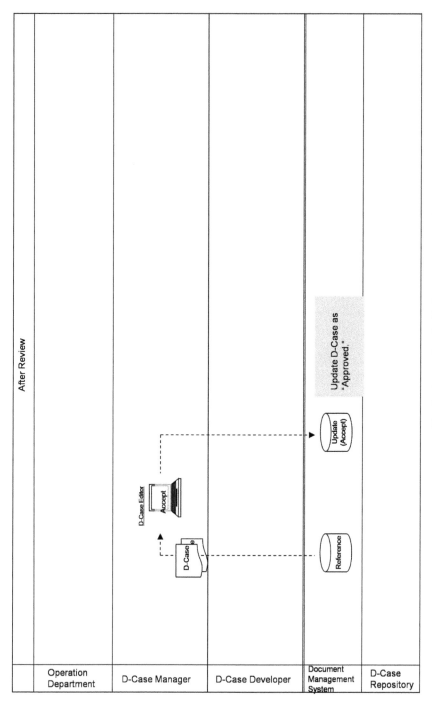

Fig. 5-6 Approving the D-Case.

(3) Implementation of D-Case Document Control

We implemented D-Case document control functionality on the basis of the workflow described above and have made it freely available. It can be used to perform the following:

- Linking to D-Case nodes of documents within the Document Management System as well as local work PCs and uploading to the Document Management System
- Detecting changes in documents linked to D-Cases and checking their version history
- Uploading D-Cases to the Document Management System
- Verification of version histories as well as comparison and searching of D-Cases
- Assignment of approved status to D-Cases in the Document Management System
- Searching for D-Cases that reference the same related documents

For interfacing with the Document Management System, we adopted CMIS,[2] which is the interface standard for the content management system defined by the standards organization OASIS. However, implementation details are product-specific, and our current development work has focused on Alfresco[3]—a CMIS-compatible, open-source content management system. Figure 5-7 shows a typical D-Case document control screen snapshot.

The software tools developed as part of this project, together with their user manuals, are freely available from the D-Case website.[4] In parallel to this, D-Case Editor was released as open-source software by Fuji Xerox and the University of Electro-Communications via a separate website.[5]

REFERENCES

[1] Fujita, H., Y. Matsuno, T. Hanawa, M. Sato, S. Kato and Y. Ishikawa. 2012. DS-Bench Toolset: Tools for dependability benchmarking with simulation and assurance, IEEE/ IFIP International Conference on Dependable Systems and Networks (DSN 2012).

[2]https://www.oasis-open.org/committees/tc_home.php?wg_abbrev=cmis
[3]http://www.alfresco.com/jp
[4]http://www.dcase.jp/
[5]https://github.com/d-case/d-case_editor

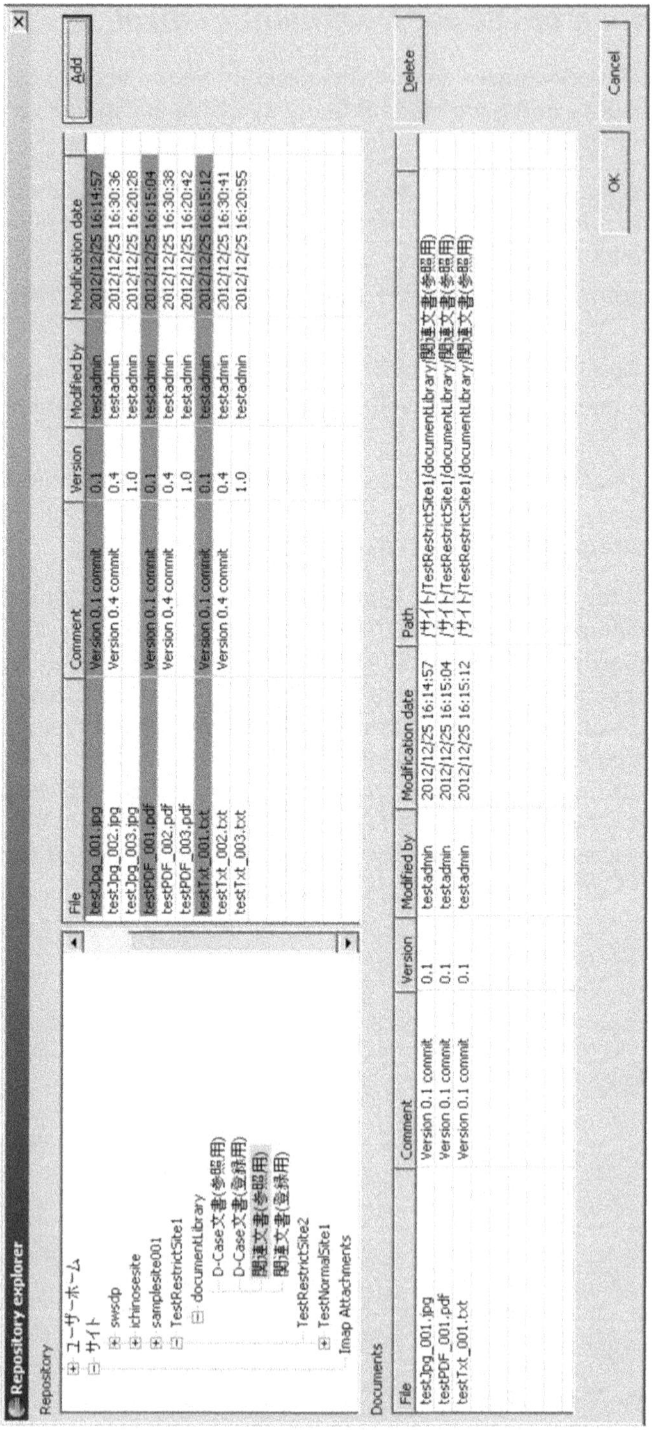

Fig. 5-7 Typical D-Case Document Control Screen Snapshot.

5.2 D-CASE WEAVER AND D-CASE STENCIL

5.2.1 D-Case Weaver

D-Case Weaver is a web-based version of D-Case Editor that provides the software's basic functions. It is compatible with *.dcase* files produced by D-Case Editor and can be used interchangeably to edit them [1, 2]. A snapshot of scrollable screen from D-Case Weaver is shown in Fig. 5-8.

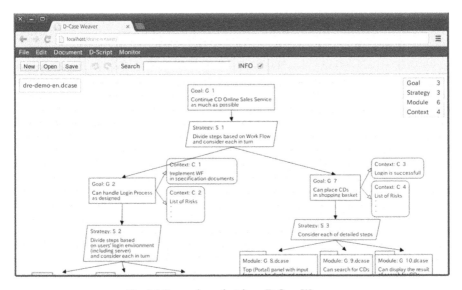

Fig. 5-8 Screen Snapshot from D-Case Weaver.

D-Case Weaver allows the following functions to be utilized within a web browser:

- Creation, addition, modification, and deletion of nodes and links in the D-Case notation for describing D-Cases
- Modularization of D-Case sections and the addition of modules to D-Cases
- Addition, modification, or deletion of D-Script information in nodes
- Creation of XML-format D-Cases that are upward compatible with those created by D-Case Editor

- Display of statistical data for each type of node in D-Case notation
- Linking with the content management system Alfresco (Community Edition) in order to manage D-Cases and related documents
- Attaching of related documents to nodes
- Setting and editing of administrator names and periods of validity

5.2.2 D-Case Stencil

D-Case Stencil is an add-in for the popular Microsoft PowerPoint® presentation software [3]. Using this add-in, you can create D-Cases for presentation purposes and small-scale practical applications. Once the D-Case Stencil add-in has been installed, a D-Case ribbon tab will be added to PowerPoint. Click on this tab to show a set of shapes for nodes and links. These shapes can be used and edited in the same way as standard PowerPoint shapes. Figure 5-9 shows a D-Case Stencil screen snapshot.

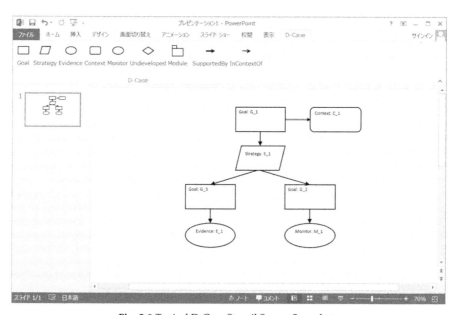

Fig. 5-9 Typical D-Case Stencil Screen Snapshot.

REFERENCES

[1]   D-Case Weaver Specifications (DEOS-FY2013-CW-01J) (in Japanese).
[2]   D-Case Weaver: http://www.jst.go.jp/crest/crest-os/tech/DCaseWeaver/index-e.html
[3]   D-Case Stencil: http://www.jst.go.jp/crest/crest-os/tech/D-CaseStencil/index.html

5.3 D-CASE & TOOLS—CURRENT STATUS AND CHALLENGES

Focusing on consensus building and accountability for dependability, the DEOS Project researchers, engineers, and collaborators from academia and industries have conducted a range of trials involving D-Case method and tool development, reference systems, and case studies. Some of the topics covered by these trials are as follows:

- Reference system containing a web server among other devices
- Web service systems for training purposes
- Sensor network systems [1]
- ET robot contests [2]
- Version control systems
- Reception robots
- Supercomputer operation manuals [3]
- Automobile engine control
- Microsatellites [4]
- Commercial broadcast system

Evaluation of processes and methods is generally carried out in line with approaches such as Quality, Cost, Delivery (QCD). We must demonstrate that D-Case can improve system quality, reduce cost, and speed up delivery. The conventional safety case has mainly been used for certification purposes, and for that reason, its application was enforceable. Work on the development of international standards to reflect the findings of the DEOS Project and the D-Case approach is under way.

The current status and challenges of D-Case research are itemized and described as follows:

(1) D-Case notation

D-Case notation has been designed on the basis of GSN to include extensions deemed necessary during the course of the DEOS Project, and a reference implementation has been realized in D-Case Editor. We are currently preparing the first concrete version of the D-Case notation specifications (as of August 2014). The GSN standard contains ambiguities [7], and notation design work is proceeding in collaboration with Tim Kelly, who originally proposed that form of notation.

(2) D-Case description guideline

One of the objectives of the description guideline described in Subsection 4.3.2 was to embed D-Case in the lifecycle of the information system. For this reason, we identified input and output documents from the various phases of this lifecycle as critical input material for developing notation. With more practical application, we will gain further experience that should prove invaluable in realizing a standard guideline. In related research, the Assurance Based Development Method [5] from the University of Virginia uses GSN to develop assurance cases for the validity of choices made in system architecture design.

(3) Linking with system monitoring and failure response functionality

We have successfully demonstrated the linking of D-Case with system monitoring and failure response functionality in development projects involving Mindstorm robots, a web-server system [6], and other applications. Further research needs to be done on the linking of D-Cases and D-Scripts. (See Chapter 8 for details.)

(4) D-Case trials

As described above, a large number of trials have been successfully conducted using D-Case. Nevertheless, a standard evaluation method for the guidelines and tools is yet to be developed. We are working toward establishing such a method through further trials.

(5) D-Case and assurance case evaluation

In parallel to the evaluation of D-Case description guidelines as stated above, evaluation of the D-Case itself—that is, quantitatively and qualitatively measuring the degree of confidence in system dependability achieved using D-Cases—is an issue of the most fundamental importance. Existing research involves, for example, quantitative evaluation of extremely simple assurance cases using Bayesian probability statistics [7] and a framework for measuring assurance case confidence based on the number of rebuttals (or defeaters) [8]; however, no approach has yet been widely adopted. It is also important in this regard that the conceptual relationships between confidence, consensus building, and accountability be clarified.

(6) D-Case tools

Until April 2010, Adelard's ASCE was, to all intents and purposes, the only tool available for creating and managing assurance cases. Now, however, D-Case Editor is available as open-source software, and other tools such as D-Case Weaver, D-Case Stencil, and AssureNote (Section 8.4) have been developed as part of the DEOS Project. In addition, Change Vision[6] is developing a GSN extension for its Astah tool.

(7) D-Case patterns

The introduction of D-Case patterns in order to increase reusability and reduce personnel dependency is extremely important from the perspective of practical application. As described in Section 4.6, D-Case patterns took their cue from Bloomfield's pattern types and the refined patterns proposed by Kelly et al. for compliance with international standards. Going forward, it will be necessary to produce guidelines showing when and where these patterns should be applied in terms of the system lifecycle or to provide standards for pattern selection. In addition, the efficacy of these patterns must be measured.

REFERENCES

[1] Nakazawa, J., Y. Matsuno and H. Tokuda. 2012. A Ubiquitous Sensor Network Management Tool Using D-Case, Abstracts of IEICE Transactions on Communications (Japanese Edition), Special Section on System Developments Supporting Ubiquitous Sensor Networks, J95-B(11) (in Japanese).

[2] Ueno, H. and Y. Matsuno. 2013. D-Case Notation Case Study for ET Robot Contest, 2013 Software Symposium in Gifu, July 2013 (in Japanese).

[3] Takama, S., V. Patu, Y. Matsuno and S. Yamamoto. 2012. A Proposal on a Method for Reviewing Operation Manuals of Supercomputer, WOSD 2012, ISSRE Workshop 2012, pp. 305–306.

[4] Tanaka, K., Y. Matsuno, Y. Nakabo, S. Shirasaka and S. Nakasuka. 2012. Toward strategic development of hodoyoshi microsatellite using assurance cases. In Proc. of International Astronautical Federation (IAC 2012).

[5] Graydon, P., J. Knight and E. Strunk. 2007. Assurance Based Development of Critical Systems, DSN 2007, pp. 347–357.

[6] Matsuno, Y. and S. Yamamoto. 2013. Consensus Building and In-operation Assurance for Service Dependability, Journal of Wireless Mobile Networks, Ubiquitous Computing, and Dependable Applications, Vol. 4-1, pp. 118–134.

[7] The GSN Work Group. 2011. GSN Community Standard Version 1.

[8] Bloomfield, R., B. Littlewood and D. Wright. 2007. Confidence: Its Role in Dependability Cases for Risk Assessment. DSN 2007, pp. 338–346.

[9] Weinstock, C., J. Goodenough and A. Klein. 2013. Measuring Assurance Case Confidence Using Baconian Probabilities, ASSURE 2013, May 2013.

[6]http://www.change-vision.com/index_en.html

6

D-Case Integrity Checking Tool and Formal Assurance Case

A D-Case that argues for the dependability of a complex, massive system is itself a complex, massive system of documents. It must be developed and updated by a large number of people, each with different responsibility for different parts of the D-Case. For the D-Case approach to complex systems to be effective, it is therefore crucial to develop computer-assisted techniques for checking and assuring the integrity of D-Case documents. Conventional review processes where team members read through the documents are not sufficient.

Integrity checking is a key part not only of D-Case development but also of maintenance. For example, special care must be taken not to lose the overall integrity of a D-Case when small portions thereof are modified to accommodate changes in an information system, its environment, or its requirements. In this chapter, we first explain how the *formal assurance case* [1] approach can be used to automate integrity checking, and we then introduce D-Case in Agda [2, 3]—a D-Case integrity checking tool based on that method.

Approaches such as the D-Case, GSN [4, 5], CAE [6], and the Toulmin Model [7] formulate the structure of assurance arguments in terms of different types of argument elements (such as goal nodes and strategy nodes) and the relationships among these elements. That said, the integrity of these arguments cannot be checked using element types alone. We need to examine the content of these elements, clarify the ontology that serves as the basis of arguments, and set that as the criteria for integrity checking. In this respect, such an ontology sets forth, by explicitly specifying words and their meanings, what objects constitute the system of interest and environment; what properties and relations are to be considered for requirements, constraints, and assumptions; what counts as verification of those properties and relations; and so forth.

D-Case, GSN, CAE, and other similar approaches do not definitively formulate an ontological definition as a structural element of the argument. Some parts of ontologies may be described in natural language in various different documents, while other parts are left implicit and unwritten, to be surmised using readers' knowledge (Fig. 6-1a). This means that the integrity of a case must be checked in comprehensive reviews done by human reviewers. Different reviewers will likely employ different evaluation criteria for integrity, but it is very difficult even to recognize such disparity. Formal formulation of an ontology that forms the basis for integrity checking is indispensable for automated integrity checking.

A formal assurance case consists of the *theory part* and the *reasoning part*. The theory part defines the ontology as a formal theory in a formal logical system. The reasoning part provides the assurance argument as a formal proof in the formal theory. Integrity checking is thus reduced to automatic syntactic checking of the reasoning part for its well-formedness as a formal proof in the formal theory (Fig. 6-1b).

The theory part provides a concrete interface through which readers and computers access the informal ontology held in the mind of the writer

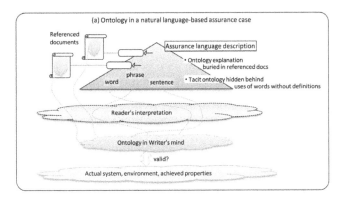

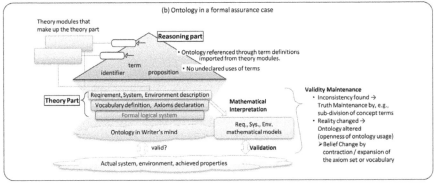

Fig. 6-1 Assurance Case and Ontology.

of the assurance case. The theory part can be subjected to validation via comparison of its mathematical models against, for example, the actual system, while the informal ontology held in the writer's mind can never be directly validated. Moreover, the explicitly described theory part is a manipulable object, amenable to operations for the maintenance of validity. For instance, when a new fact comes to light and brings an inconsistency to the current ontology in use, some truth maintenance algorithm may be invoked to recover consistency by, for example, sub-dividing a concept term into several terms that denote more refined concepts. Also, when the reality changes itself, the changed objects and concepts in the new reality often have not been conceived as being potentially possible. In fact, it is often the case that the objects and concepts before the change are often recognized as such only after, and because of, the change in reality. Thus, the ontology in use before the change would not provide the necessary vocabulary for recognizing and expressing that "*that* thing (fact) has changed into *this* thing (fact)". To alter the ontology and maintain its validity with respect to the changing reality, operations on theories as studied in the field of *belief change* may be employed—for example, contraction and expansion of axiom sets.

Assurance of open systems requires a notion of "open" ontologies. Such an ontology would combine a definite description of terms and their meaning, which is to be used on the premise that it will change, and a mechanism for maintaining the validity of that description. Establishing such a mechanism for validity maintenance is a major step in achieving the DEOS ideal: despite recognizing that it can never be known how things will change, equip the system in advance with the ability to exert best effort and perform when a change does occur. Traditional formal methods tend to focus on verification within a fixed theory that serves as *the* frame of reference. A formal assurance case, having the theory part as manipulable data, is a formulation that anticipates changes in that frame of reference.

The integrity checking tool, D-Case in Agda, is a tool that implements a part of the formal assurance case methodology described above. It makes it possible to develop and check formal assurance cases using the Agda programming language—a language based on the *propositions-as-types, proofs-as-programs* correspondence. The theory part of the assurance case is described as a collection of libraries providing definitions of types and functions; the reasoning part is described as a program (i.e., a proof) using those types and functions; and integrity checking is reduced to type checking of the program. Use of programming language features allows for development of larger, more complex formal theories and proofs than is practically doable with traditional notations used in mathematical logic. D-Case in Agda can convert arguments as Agda programs to/from those in the graphical D-Case form, enabling a combination of automated integrity checking and content-based review by domain experts.

6.1 BENEFITS OF FORMAL ASSURANCE CASES

(1) Assurance communication

Formal assurance cases provide for more accurate assurance-related communication between stakeholders.

With assurance cases described using GSN and other similar techniques, the boundary between what constitutes the basis of reasoning and what the reader must interpret and judge using expert knowledge may not be clear. If the way in which a goal is decomposed into sub-goals is not obvious, the reader must consider a number of different possibilities:

1. The reader is lacking the expert knowledge implicitly expected of him/her.
2. The writer of the case is implicitly stipulating that the goal and sub-goals have the meaning that makes the decomposition correct. For example, when certain sub-goals purportedly support a goal that the system is dependable, the reader might need to assume that the writer is defining the meaning of "dependable" to be the conjunction of the sub-goals.
3. The reader needs to work harder to understand how the decomposition is substantiated by the information presented.
4. The writer made an error in decomposing the goals.

With a formal assurance case, the theory on which the reasoning is based and the argument itself are defined explicitly and separately from one another, and automated checking ensures that the argument is correct with respect to the theory. Possibilities 1, 2, and 4 above can thus be eliminated using this method. For Possibility 3, meanwhile, the reader should examine how the decomposition is derived from the theory and convince himself/herself of its appropriateness, or alternatively, identify the parts of the theory providing justification for the unacceptable decomposition and raise an objection concerning the appropriateness of those parts.

The mathematical interpretation of the descriptions of goals and the like is uniquely determined once that of that basic vocabulary provided by the theory part of the formal assurance case is defined. Regardless of how the vocabulary is interpreted, as long as the axioms of the theory part are satisfied, the goal will be true under this interpretation. When evaluating the validity of a case, each stakeholder examines (1) whether the results of interpretation of the descriptions conform to the original intent, (2) whether the interpretation of the axioms holds in reality, and more generally, (3) whether the formal theory appropriately models the actual information system and its environment. These pieces of information for evaluation are more definite than in the case of informal assurance cases, and they can be shared by stakeholders. This also makes it easier to identify the cause of any disagreement among stakeholders concerning the evaluation.

Formality of a case description has nothing to do with how detailed it must be. By enabling communication based on rigorous interpretation, the formal assurance case can be beneficial regardless of its level of detail.

(2) Automated integrity checking

Various kinds of integrity checking can be performed on a formal assurance case. Much of the checking carried out during reviews that does not require expert judgments can be replaced by automated checking of whether the description of reasoning constitutes a well-formed formal proof in a formal theory. Since the description of the theory is also formal, it too can be checked for internal inconsistencies such as duplicate definitions. Appropriateness of the content of the theory must be judged in the usual processes of manual reviews and consensus building.

Argument trees in GSN and other similar graphical notations can be checked for integrity based only on the types of argument elements used and the way the elements are interconnected. Manual review is needed even to determine whether the description contained in a goal node actually is a proposition (goal), whether the strategy is suitable for decomposing the goal, and so forth. With formal assurance cases, computers can examine the structure of the internal description of argument elements, making possible more detailed integrity checking.

The role played by formal languages in the development of formal assurance cases is analogous to that of typed languages in programming. Programming errors found by type checking are mostly simple ones, and avoiding type errors is only one of the difficulties in programming. However, finding all type errors by hand is very difficult, and type checking has become an indispensable feature for programming by freeing programmers from such worries and by letting them concentrate on more important issues. In the same way, automated integrity checking allows reviewers to better apply their expert judgment in evaluating the content of assurance cases.

(3) Change management

Formal assurance cases provide for more systematic, principled automation of change management than informal ones, since they have more detailed structures that make the meaning of argument elements and the relationships among them explicit to computers. Automation would include version control, traceability analysis, cause and effect analysis, change impact analysis, and so forth.

6.2 FORMAL ASSURANCE CASE

A formal assurance case comprises a theory part and a reasoning part. The theory part provides a formal theory that defines the vocabulary to be used in describing the reasoning part and the mode of use thereof. The reasoning part provides an argument as a formal proof (that is, a proof written in a format that can be processed by machines) in the formal theory. In the following, we first describe the basic concept of a formal assurance case in terms of a usual formulation of formal logic. We then explain how formal assurance cases can be written as Agda programs.

(1) Basic concept

(1) Reasoning part

The structure of an argument in graphical notations such as D-Case and GSN is essentially the same as that of a formal proof in the natural deduction style (one of the standard formats for formal proofs in the field of mathematical logic). In general terms, the correspondence between these two can be summarized as follows:

- Goals are propositions.
- Strategies are inference rules (including derived inference rules that use axioms).
- To decompose a goal by a strategy into sub-goals is to infer a conclusion by an inference rule from premises.
- To support a goal by evidence is to show that a proposition is an axiom by an axiom schema.
- Formal proof of a proposition *A* has one of the following formats:
  - Application of the inference rule whose conclusion is *A* to formal proofs of the premises of the rule.
  - Application of the axiom schema with *A* as the conclusion.

Accordingly, checking whether the reasoning part of a formal assurance case constitutes a formal proof involves checking whether or not the strategies and evidence as defined in the theory part are combined in the prescribed manner.

Much of the research in the field of the assurance case emphasizes the point that *an argument is not proof*, and research, clearly acknowledging the above-described correspondence has only recently gotten under way [8, 9, 3]. Stating that an argument is not proof usually means that an argument is not *pure logical proof derivable from universally true axioms*. While this is true, it does not imply that all checking of the validity of arguments must be carried out manually. The theory part of the formal assurance case can make explicit special circumstances on a case-specific basis. As a result, division of work becomes possible between experts who validate the theory part and machines that check the integrity of the reasoning part.

(2) Theory part

The purpose of the theory part of the formal assurance case is to declare the axioms used in the reasoning part, including those for creating new inference rules as derived inference rules. For this, it also specifies what propositions there are, what predicates can be used to form propositions, and what objects there are. This is done by declaring vocabulary and the mode of use thereof.

The theory part comprises a formal logical system selected to serve as the basis and a definition of a formal theory in that system. For example, we take here the many-sorted first order predicate logic as the base formal logic. Then, a formal theory therein is given by specifying the following:

- Sort symbols (i.e., symbols that represent sorts (types) of objects)
- Constant symbols and their sort information, and function symbols with sort information for their arguments and results (i.e., symbols that represent basic objects and the basic constructions thereof; they form terms for describing complex objects)
- Proposition symbols, predicate symbols, and sort information for their arguments (i.e., symbols that represent the relationships between the basic concepts and objects; together with constant symbols, function symbols, and logical connectives, they form terms for describing propositions)
- Axioms (i.e., terms for describing a proposition as a basic assumption)

These symbols may be interpreted in any way as long as all the axioms are true under that interpretation. Generally speaking, the number of axioms may be infinite, and they are given by axiom schemata—i.e. (parameterized) inference rules with no premises.

The fact that both the theory part and the reasoning part are formal does not mean that deep logical analysis is necessary. Indeed, given any argument tree in D-Case or GSN notation, it is easy to define a formal theory in which the argument is a formal proof. One can take each goal as a primitive proposition represented by a proposition symbol, and each item of evidence or strategy as a derived inference rule derived from a purpose-made, ad hoc axiom. While nothing can be gained from integrity checking of the argument with respect to such a theory, this formulation makes explicit that all strategies are postulated ones without any backing. The benefit of the formal approach is that, regardless of the depth of analysis, the assumptions are clear to both human and machine.

(3) Features required of a formal logical system

The concept of a formal logical system is fundamental in mathematical logic and is the guiding principle behind the formal assurance case. However, the conventional formulation of this type of system lacks the following two mechanisms for defining complex and large formal theories needed for formal assurance cases.

- *Definition mechanism*: In order to organize logical analysis, it must be possible to introduce new concepts and abstractions on top of a primitive notion represented by basic symbols. For this, it must be possible to make and use definitions that assign new names to complex terms, propositions, derived inference rules, and the like constructed from previously declared or defined terms.

- *Declaration mechanism*: Declaration of basic symbols and axioms must have a machine-processable, formal syntax.

In the usual presentation of formal theories on paper, necessary definitions and declarations are shown at the level of running text in a manner suited to human readers. When describing formal assurance cases, however, the declaration and definition mechanisms must be clearly formulated so that machines can also process them. Notations such as D-Case and GSN use context nodes to refer to definitions and declarations given in other documents. However, the referenced definitions and declarations are not themselves formulated, and therefore, it is not possible to check whether they are consistent. Formulation of formal assurance cases makes it possible to check the consistency of definitions and declarations.

(2) Describing formal assurance cases in Agda programming language

(1) Agda language and the paradigm of propositions-as-types, proofs-as-programs

Agda is a general-purpose, functional programming language with dependent types based on constructive type theory, which is a foundational theory of mathematics [10]. Agda is also a description language for propositions and proofs in higher order intuitionistic logic via the principle of *propositions-as-types, proofs-as-programs* (Brouwer-Heyting-Kolmogorov interpretation, Curry-Howard correspondence). Roughly speaking, the correspondence between proofs and programs under this principle is as follows:

- A proposition is a data type. This data type specifies what data counts as direct evidence that shows the proposition to be true.

- An inference rule is a function that constructs direct evidence data for its conclusion from direct evidence data for its premises.

- An axiom schema is a constant or function that constructs direct evidence data for the axiom proposition.

- Formal proofs are programs that combine the above constants and functions and construct direct evidence data for their conclusion.

Combining this with the above correspondence between formal proofs and arguments in D-Case and GSN notations, we get a *goals-as-types, arguments-as-programs* correspondence. More specifically, this means that:

- Integrity checking of an argument corresponds to type checking thereof as a program;
- A formal theory corresponds to a set of library modules declaring and defining the types, functions, and constants used in the definition of an argument as a program; and
- Context nodes correspond to *open declarations* (described below) that allow the types, functions, and constants declared and defined in the library modules to be used in arguments as programs.

Agda is but one language suited to the description of formal assurance cases, and other languages with similar features could also be used. The necessary features include a powerful type system that allows propositions to be expressed as types, static type checking that guarantees termination, and functionality for abstraction and modularization.

(2) Agda description of theory part

For the purpose of explanation, Fig. 6-2 shows a simplified version of an example goal structure from the GSN Community Standard [4].

The Agda code for the corresponding formal assurance case can be found at the end of this chapter. The above argument tree claims that the control system under consideration is acceptably safe, and this goal is supported by the facts that all identified hazards are appropriately dealt with and the software development process has been appropriate.

The theory part must define the vocabulary for describing the concepts and objects appearing in this argument tree. In specific terms, this includes the following.

(a) Vocabulary for describing objects and their types

The declaration

```
postulate
  Control-System-Type : Set
  Control-System : Control-System-Type
```

introduces vocabulary identifying the type `Control-System-Type` for control systems in general and identifies `Control-System` as the specific control system under consideration. This `postulate` declaration posits as an axiom that type and object with these names exist but no properties other than existence are assumed here.

Identified hazards are introduced by declaring the enumeration type whose elements represent each and all of them.

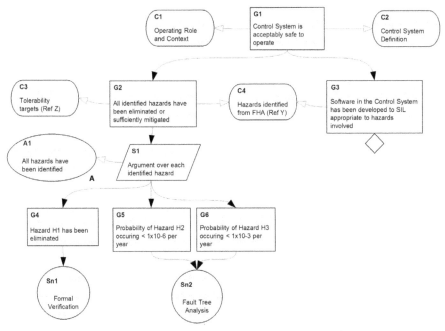

Fig. 6-2 Example Description in GSN Community Standard.

```
data Identified-Hazards : Set where
    H1 H2 H3 : Identified-Hazards
```

(b) Vocabulary for describing goals and axioms

The following declaration introduces the proposition that *Software in the Control System has been developed to SIL appropriate to hazards involved* as a primitive proposition.

```
postulate
    Software-has-been-developed-to-appropriate-SIL : Set
```

The predicate *X is acceptably safe* on control systems in general is introduced as a predicate (propositional function) with a control system as an argument.

```
postulate
    Acceptably-safe-to-operate : Control-System-Type → Set
```

Applying this predicate to `Control-System`, the top-goal proposition that *the control system is acceptably safe to operate* becomes the function application `Acceptably-safe-to-operate  Control-System`. By defining the infix operator `is` for function application, the same proposition can be written `Control-System is Acceptably-safe-to-operate`.

The following declaration introduces the axiom that this top-goal proposition is implied by sub-goals. *All identified hazards have been eliminated or sufficiently mitigated* and *Software in the Control System has been developed to SIL appropriate to hazards involved*. The name `argument-over-product-and-process-aspects` is that of the axiom schema (inference rule with no premises) whose conclusion is this implication proposition. It is at the same time that of the strategy (derived inference rule) given by this implication proposition and the inference rule of *modus ponens* (the two amount to the same thing under the proofs-as-programs interpretation).

```
postulate
   argument-over-product-and-process-aspects :
      (∀ h → h is Eliminated Or Sufficiently-mitigated) →
      Software-has-been-developed-to-appropriate-SIL →
      Control-System is Acceptably-safe-to-operate
```

A reviewer would certainly take issue with the acceptance of such an ad hoc axiom. The important point here is that the formal nature of this approach forces the writer to openly admit that the axiom has no backing in the form of logical analysis (otherwise, a definition would have been used instead of `postulate`).

The predicate *X has been sufficiently mitigated* on Identified-Hazards is not introduced as `postulate` but defined using a previously declared, more primitive concept.

```
Sufficiently-mitigated : Identified-Hazards → Set
Sufficiently-mitigated h =
   Probability-of-Hazard h < mitigation-target of h
```

(c) Referencing evidence

From the formal perspective, assigning a name to evidence for a goal so that it may be referenced within the Agda code is the same as introducing and naming an axiom schema with that goal as its conclusion.

```
postulate
   Formal-Verification : H1 is Eliminated
```

However, since there are significant differences between reviewing evidence and reviewing the suitability of axioms that form the basis of reasoning, names for evidence are declared separately from those for axioms.

(d) Defining contexts

Declarations and definitions of the above vocabulary and axioms are grouped into multiple modules. Context nodes appearing in an argument tree correspond to an `open M` declaration, which opens a specific module M and makes it possible to reference the names declared and defined inside it. (Modules are used to manage the namespace. The names declared and defined inside one cannot be referenced directly from outside it. In order to do so, the user of the module must explicitly open it, making the dependency to the module explicit.)

(3) Agda description of reasoning part

With the *proofs-as-programs* correspondence, any Agda expression having the type corresponding to a concluding proposition can be considered a formal proof. However, Agda code in the reasoning part uses the following specific style that makes explicit the structure of Agda expressions as argument trees in D-Case or GSN notation.

A case is an expression that has either of the following two forms:

- <goal> by <argument>: This corresponds to an argument tree whose top node is the type <goal>. The infix operator by is the identity function on the type <goal>. If the value of <argument> is of the type <goal>, the entire expression takes this value; if not, a type error would result.
- let open <module> in <case>: A context is attached to the top goal of <case>, and names declared and defined inside <module> can be referenced within <case>. The entire form takes the value of <case>.

The format for the expression <argument> can be one of the following:

- <strategy> • <case$_1$>... • <case$_n$>: This corresponds to the tree structure whose root node is the function <strategy> and whose immediate sub-trees are the <case> entities. The infix operator • is the function application, and if the type expected of each ith argument by the function <strategy> matches that of the <case$_i$> (i.e., the top goal type of <case$_i$>), the entire expression takes the value of the function application having the function <strategy>'s result type; if not, a type error would result.
- <evidence>: This corresponds to an evidence node. It may be any well-formed expression as long as its type matches the parent goal type indicated by <goal> by...; if not, a type error would result.
- let open <module> in <argument>: A context is attached to the top node of <argument>, and names declared and defined inside <module> can be referenced within <argument>. The form takes the <argument> value.

With this correspondence, the definition body of the function main shown at the end of this chapter directly corresponds to the GSN argument tree shown in Fig. 6-2.

In this particular example, the theory part reflects only a very shallow and abstract analysis of the argument, and integrity checking with respect to this does not increase much confidence in the argument. Even so, when the theory and reasoning parts are updated by different writers, if, for example, the argument were not updated after an addition of elements to `Identified-Hazards`, or, if the `mitigation-target` for each hazard were changed but the same evidence was used as it is, these problems would be detected as type errors when the entire code is re-type checked, with the location of errors pointing to the parts where fixes must be made.

(3) Extension to formal D-Case

The concept of *arguments-as-Agda programs* provides a definite formulation and meaning to the logical aspect of D-Case that supports integrity checking, enhancing the following characteristics of D-Case as an extension of the informal assurance case.

(1) D-Case patterns and modules

Patterns can be defined as functions that compute from parameter values the expressions in the style described above.[1] An argument tree containing an instance of a pattern corresponds to a program containing a function call expression to the pattern function (after inlining the call by partial evaluation). In addition to familiar data such as numeric values or strings, the parameters of pattern functions may be goals (types), strategies (functions), argument sub-trees, predicates, proofs that parameter values satisfy some constraints, and so forth. Pattern function definitions are ordinary Agda function definitions. Definition bodies are type checked based on the specified typing of parameters and results, guaranteeing that any well-typed call to the pattern function will result in a consistent argument. There are no restrictions on how the result is computed—multiplicity and selection as seen in GSN patterns, or alternatively, terminating looping (recursion) and the like can be systematically programmed in Agda.

Modularization benefits argument construction as it does programming: a large argument may be constructed as a combination of components each of which can be managed with a degree of independence. The functionality of GSN modules and contracts can be realized using Agda's module mechanism. The functionality provided by this module system includes hierarchical definitions of parameterized modules, split type checking on an individual module file basis, importing of modules defined in other files, managing the visibility of names declared or defined inside the module (i.e., public to other modules, private, or abstract (showing a name and its type but not its definition body)) and renaming.

---

[1]For simplicity, the explanation given here glosses over the distinction between Agda expressions and their values. In practice, processing based on the forms of expressions is suitably performed through partial evaluation of programs and/or internalization (deep embedding) of relevant expressions as Agda data.

Current proposals concerning the formulation of argument patterns and modules often show ambiguous, ad hoc or awkward handling of issues such as free and bound variables, the scope of declarations and definitions, and terminating recursion. From the programming language perspective, these are problems that have principled solutions and there is no need to develop new ones from scratch.

(2) External nodes and inter-dependability cases

External nodes correspond to references to names defined in other modules. Suppose that, in a D-Case argument tree for a system A, a goal GA is supported by an external node linked to the D-Case for an external system B. Call the theory part module of system A's formal D-Case TA and the reasoning part module RA; similarly, call the corresponding modules of system B's D-Case TB and RB, respectively. In the simplest case, module RA imports RB and shows GA by a <case> expression GA by nB, using some name nB defined in RB of some <argument> expression. A type error will result if the goal GA and the type GB of nB do not match, but it is important to note that GA is described using the vocabulary defined in TA, and GB using that in TB.

The following conditions must be satisfied for GA and GB to match: (i) TA and TB have imported the same library module TC; (ii) when the definitions of the terms used in the descriptions of GA and GB are expanded, they are described using TC terms only; and (iii) the two descriptions match after expansion. Furthermore, unless (iv) TB does not add axioms concerning TC terms, the interpretation of GA changes when RA imports RB (hence TB indirectly) and makes reference to nB.[2] It is preferable to describe the interfacing parts of system A's and B's D-Cases using a common vocabulary from the outset, and developing vocabulary libraries to make this possible is one important approach in this regard. Unfortunately, it is more often the case that the above conditions cannot be met as systems A and B are developed and operated by independent groups of stakeholders.

In this case, an additional reasoning that GA can be argued for using TB and RB, and the theory that backs that reasoning, will be required. These are provided by the reasoning part module RD and the theory part module TD of the inter-dependability case [see d(A,B) and d(A,C) from Section 4.5]. TD imports TA and TB, and it adds axioms for interpreting TB vocabulary using TA vocabulary. RD imports RB, and using the names defined in RB and the axioms added by TD,[3] it defines the name nD of an <argument> expression of type GA. In this case, use of the external node in the argument tree of system A corresponds to RA's importing RD and showing GA by the <case> expression GA by nD.

It is very common for two D-Cases to use the same terms with different meanings. Mixing up of such terms is a major cause of inconsistency when two

---

[2]Situations where this is accepted are quite possible.
[3]The direct use of TA and TB terms is also permitted.

D-Cases are combined. However, formulation of external nodes as described above and type checking can guarantee that this type of confusion does not occur. Two identifiers (names) declared or defined in different modules are different even if they are the same as lexical tokens, and mixed usage would be detected as a type error or ambiguous name error. When reviewing natural language descriptions, it is extremely difficult to keep the two separate and prevent a mix-up.

(3) Monitor nodes and action nodes

A D-Case argument containing monitor nodes and action nodes corresponds to an "IO *A*" type program that exchanges data with the outside world, changes the world to make true the top-goal proposition *A*, and returns direct evidence data for *A* as results.

(a) IO types

We should start by briefly explaining how the Agda language handles input and output. For a type *A*, the value of the type IO *A* is generally a command that, when executed, returns a result of type *A*. For example, types for the getLine command that gets input from the user and returns a string, and for the putStr command that outputs a string passed as an argument are as follows.

```
getLine : IO String
putStr : String → IO Unit
```
(type Unit is the type of a dummy result of no particular interest)

More complex commands are constructed by combining these basic ones.

The command m >>= f is constructed from a command m and a function f that returns a command as a result. When it is executed, the command m is first executed and its result a is obtained. Then, the command f a is executed and its result is returned as the result of executing m >>= f. Here, f is a function that calculates a command from an argument value. When f is the function λ x → c (c is an expression for a command containing the argument x), it is also written as do x ← m then c. The command return a returns the value of the expression a without any input or output.

Commands are a kind of value, and an Agda program that performs input and output is a program that computes a complex command using the above operators. Type checking guarantees that a program of type IO *A* computes a command that returns a result of type *A* when executed.

Basic commands are implemented using different languages, and they are associated with Agda command names (such as putStr) using Agda's

Foreign Function Interface (FFI).[4] The typing of a command name is one part of the specification that the command's implementation must satisfy (argument and result types). The fact that it does is taken as an axiom for Agda proofs. To explicitly reason over possible failures of command execution, typing of the result may be changed to IO (*A* or error) or the like. Even in this case, it is taken as an axiom that determination of successful execution or failure itself does not fail.

(b) Monitor nodes and action nodes in natural language-based D-Case

Monitor nodes in D-Case arguments constitute dynamic evidence based on the state of the operating system and its environment. In natural language-based D-Case, at least two modes of use can be distinguished:

1. Usage as evidence for a goal such as *the state of the system and the environment is in an expected range*: In this case, the goal is considered to be satisfied without having to develop an argument, and the monitor node is added either as a precaution or to provide live evidence to be logged. Two further sub-cases of this are:

   (i) The design of the system expects and covers exceptional situations where the goal is not met, and the D-Case contains arguments for these situations.

   (ii) Design does not cover such situations, implicitly indicating that detection of such triggers a transition to the *Change Accommodation Cycle* in accordance with the general DEOS principle.

2. Usage as evidence for a goal such as *whether the system is in a desired state or not can be determined*: Both situations are covered by the design, and the D-Case contains an argument for each.

   Mode 2 is more general as it can also express the intention of Modes 1(i) and 1(ii). When applied as it is, Mode 1(i) leads to confusion regarding the actual sense of displaying evidence to support a goal. Accordingly, we take Mode 2 as the basic usage to be formulated.

   In both Mode 1 and Mode 2, it is presumed that the success or failure of the monitoring itself is discernible, and the failure of monitoring is not covered by the design. If developing a D-Case for a design that also takes such failure into consideration, the goal must be expressed more explicitly: *[It is possible to determine [whether or not it is possible to determine [whether the system is in the desired state]]]*.

---

[4]In the current Agda implementation, only Haskell is supported as a foreign language. Implementation with other languages is possible by further using the Haskell FFI mechanism.

Action nodes are used as evidence for goals such as *the system can do <some action>* or *<some action> is taken*. These goals are often claims that the system or its environment will be in a certain state when the action is executed.

(c) Formulation of monitor nodes and action nodes

A monitor node *M* that determines whether or not a proposition *A* is holding is formulated as a command of type `IO(Dec A)`. `Dec` is an abbreviation of "Decidable", and the value of the type `Dec A` is either:

- `yes` *p*, where *p* is direct evidence data supporting *A*, or
- `no` *q*, where *q* is direct evidence data supporting ¬*A*, the negation proposition of *A*.

A function receiving the execution result of *M* does case analysis of the result and computes either an argument using *p* for situations where *A* holds or an argument using *q* for situations where ¬*A* holds, as the case may be. *M* is not limited to basic commands: it is also possible to program a monitoring command in Agda that computes a value in `Dec A` by calling a basic command that returns some numerical value and then determining *A* or ¬*A* using that value.

Action to make proposition *A* true is formulated as a command of type `IO A`. The fact that the proposition *A* actually holds after execution is taken to be a basic assumption about the command. If failures must be explicitly considered, the action should be formulated as a command of type `IO` (*A* `or error`) so that a case analysis on success or failure of the execution result is possible.

It should be noted that, whether for monitoring or for actions, the description of proposition *A* must explicitly mention time and other changeable factors. This is because the way the description is interpreted is unchangeable, irrespective of time and the execution status of the argument program. If, for example, an action that changes a state where *OK* does not hold into a state where it does is formulated as a command `recovery : ¬ OK → IO OK`, one can construct another command `magic : ¬ OK → IO A` that makes true any proposition *A* when evidence for ¬ *OK* is given. Instead, more explicit formulation such as `recovery : ∀(t : Time) → ¬ OK t → IO(OK(t +1))` is necessary to avoid such inconsistency. This kind of description quickly becomes too cumbersome in practice. Similarly to Hoare logic, a concise shorthand notation should be developed by taking state-dependent predicates, rather than propositions, as the basis for description and by implementing a framework that links the execution of IO commands with state transition.

(d) D-Case arguments as IO programs

Input/output capability makes it possible to extend the correspondence between arguments and programs and the meaning of the correspondence.

- An argument having a goal G as the top node corresponds to a program of type `IO` A.[5]
- A strategy for decomposing the goal G into sub-goals G_1 through G_n corresponds to a function that constructs a command of type `IO` G from a command of type `IO` G_i (the function itself may perform input and output).
- Evidence for goal G corresponds to a command of type `IO` G.
- Integrity checking of the argument corresponds to type checking of the program. Type checking guarantees that the program constructs a command whose execution makes the top goal true and returns direct evidence data for the top goal, provided that assumptions on basic commands taken as axioms hold true.

Static arguments without monitoring or actions are special instances of IO programs within the above framework: they are equivalent to arguments with IO in which all commands appearing have the form `return` a that performs no input or output. The execution of a static argument program was not mentioned because type checking suffices to guarantee that the top goal proposition holds: execution does not change the state of the system or environment and has no effect on the truth value of the top goal. Note that the above unnaturally restricts the structure of IO programs to the simple structure of argument trees. If the graphical notation could be extended to support, say, syntax for exception handling in programming, it would provide for clearer presentation of arguments.

D-Case arguments as IO programs provide the most straightforward manifestation of the idea that the stakeholders' agreement is (would be) satisfied by operating the system in accordance with the D-Case description.

6.3 FORMAL D-CASE AND SYSTEM OPENNESS

(1) Common misconceptions

In the formal D-Case for a system at a point in time, its formal theory is a formulation of the best understanding of the system and environment at that time. The formal D-Case sets forth a formal argument that is closed within

---

[5]In order to include time-related information in claim G as described above, it will be necessary to consider, for example, a program of type $\forall$(`t` : `Time`)$\rightarrow$`IO(G(t))` that takes the current time t as an argument.

the models of that theory.[6] If the system and environment change and are no longer a model of the theory, the argument becomes unfounded. However, the claim that this renders the formal D-Case less suited to open systems than a D-Case described using natural language is often based on the following misconceptions:

- "For the very reason that a natural language admits ambiguity, natural language-based D-Case arguments allows for re-interpretation in line with changes in the system or its environment": This misconception is based on the mistaken notion that defining a formal theory equates to the definition of one fixed, unique model. Defining a formal theory does not set an interpretation, but fixes and clarifies the terms that are open to (re-)interpretation and the range of possible interpretations that does not jeopardize the integrity of arguments.

- "A fixed formal theory cannot be used for arguments about systems that undergo change, including foreseeable change": This misperception often comes from a failed attempt to formulate natural-language descriptions that have time-dependent or state-dependent meanings (e.g., "free space is available on the hard disk") without taking those dependencies into consideration.

- "The specifications of an open system can never be complete, so formal descriptions either cannot be used or would be meaningless": This false impression is caused by the mistaken notion that the significance of formal description lies in the development of a complete and sound formal theory for the actual system. Firstly, there is no need for the system's formal theory to be so complete that all propositions concerning the system can be proved.[7] It is sufficient to be able to prove relevant propositions from axioms that have convincing justification. Secondly, the soundness per se of the system's formal theory is not the purpose of defining it. No matter how detailed the axioms, one can never fully model the actual system and it is impossible to ensure the soundness (any proposition that is provable in the theory is true in the actual system). Even so, the implication from the axioms to the conclusion is actually true, so causes of a false conclusion can be traced back to false axioms, for which improvements can be made. The significance of having formal description is that it provides more reliable staging points than natural language description when iteratively improving the precision of approximations in this way.

---

[6] A model $M=<S,I>$ of a formal theory T consists of a structure S of meaning and an interpretation I that assigns elements of S to symbols of T such that I interprets all axioms of T as true sentences. S itself may be called a model leaving I implicit. This technical sense of "model" is different from its more usual sense (i.e., a simplification or abstraction) as in *consider the theory T as a model of the system.*

[7] Complete descriptions are also impossible in principle in systems that incorporate natural number theory (as per the incompleteness theorem).

(2) Formulation of rebuttals

Unforeseen changes in an open system rarely result in the system simply taking on new properties; instead, it is much more likely that certain properties cease to hold at the same time. Correspondingly, changes to the system's D-Case need to incorporate new assumptions that conflict with those already there.

With a natural language-based D-Case lacking a theory part, the pre-change and post-change arguments are both based on vocabulary with vague interpretation in the mind of the reader at the respective points in time. A situation where knowledge (i.e., vocabulary and its interpretation) increases with contradiction cannot be formulated using usual logic, and it is necessary to resort to special approaches such as non-monotonic logic and paraconsistent logic.

On the other hand, a formal D-Case, which explicitly contains the theory part on which each argument is based, can formulate this type of situation in a more straightforward manner as changes to that theory part, thereby eliminating the need for special logic. For example, rebuttal of the conclusion A of an argument and resolution of the contraction that this rebuttal brings about could be seen as the types of change shown below. Here, Fig. 6-3 refers to the simplest well-known example of non-monotonic reasoning—*the penguin Tweety is a bird but cannot fly.*

Thesis (T, A, p): Advocate for the thesis claims that the proposition A holds and gives the theory T and the proof p of A in T. In this specific example, A is Axiom 1 (*All birds fly*) of T. (T is a theory comprising this axiom and declarations of the predicate symbols *bird* and *flies*.)

Antithesis (T_C, T_R, q): Rebutter[8] who claims that $\neg A$ (the negation of A) is the case first examines the proof p of A in T and divides the theory T into the part to which he or she cannot consent and the remainder. The remainder becomes the sub-theory T_C of T that is common to Advocate and Rebutter. (For brevity, we assume that T_C includes the declaration of the basic vocabulary needed for describing A.) Next, Rebutter extends T_C to the theory T_R as the basis for claiming $\neg A$ and presents a proof q of $\neg A$. The sum of T and T_R is an inconsistent theory and any proposition can be proved in it; however, it should be noted that q is not a meaningless proof in this inconsistent theory.[9] In this specific example, T_C comprises only declarations of the predicate symbols, and T_R adds to this the class *Penguin* of flightless birds and the instance *Tweety* thereof.

---

[8]Advocate and Rebutter need not be different people. Whenever counterevidence to A is identified, Advocate must rebut an argument he or she made in the past.

[9]It is conceivable that at the first part of rebuttal, Rebutter finds nothing disagreeable and $T = T_C = T_R$. In this case, both Advocate and Rebutter have accepted an inconsistent theory T, and inconsistencies therein must be removed first.

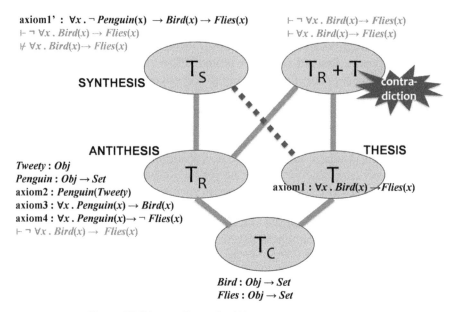

axiom1' : $\forall x . \neg Penguin(x) \rightarrow Bird(x) \rightarrow Flies(x)$
$\vdash \neg \forall x . Bird(x) \rightarrow Flies(x)$
$\nvdash \forall x . Bird(x) \rightarrow Flies(x)$

$\vdash \neg \forall x . Bird(x) \rightarrow Flies(x)$
$\vdash \forall x . Bird(x) \rightarrow Flies(x)$

SYNTHESIS T_S

$T_R + T$ contra-diction

ANTITHESIS   THESIS

Tweety : Obj
Penguin : Obj $\rightarrow$ Set
axiom2 : Penguin(Tweety)
axiom3 : $\forall x . Penguin(x) \rightarrow Bird(x)$
axiom4 : $\forall x . Penguin(x) \rightarrow \neg Flies(x)$
$\vdash \neg \forall x . Bird(x) \rightarrow Flies(x)$

T_R

T
axiom1 : $\forall x . Bird(x) \rightarrow Flies(x)$

T_C

Bird : Obj $\rightarrow$ Set
Flies : Obj $\rightarrow$ Set

Fig. 6-3 Well-known Example of Non-monotonic Reasoning.

Synthesis (T_S, A', p'): Here we consider the situation where the antithesis arises from new counterevidence to A and Advocate agrees with T_R. In order to regain as much as possible of the meaning content of A in T, Advocate (or both Advocate and Rebutter) presents the theory T_S that extends T_R, a new proposition A' to substitute for A, and the proof p' of A' in T_S to remove the contradiction in $T_R + T$. Because T_S necessarily proves $\neg A$ for it to be an extension of T_R, it must not prove A for it to be consistent. In particular, T_S cannot be an extension of T. In this specific example, A (*All birds fly*) that is negated by T_R is replaced by A' (*All non-penguin birds fly*; Axiom 1').

Research remains to be done to clarify the sense of A' "regain[ing] as much as possible" of the meaning content of A and specific methods for constructing T_S and A'. In the example shown above, A' (Axiom 1') is weaker than *All birds other than Tweety fly* and stronger than *All flying birds fly*, but it must be possible to argue the most desirable solution.

Rebuttal of D-Case arising from requirement changes and unforeseen failure, as well as the corresponding change accommodation, share common traits with this simple example. In each case, formal D-Cases with formulation of pre-change and post-change theory parts as manipulable objects are indispensable even for discussing issues such as the above.

(3) D-Case Validation with assurance metacase

Because a formal D-Case has a definite format and meaning as data, it can be subjected to computer-assisted verification that determines whether or not it has certain desired properties. Taking a desirable property of a formal D-Case to be the fact that it addresses issues of openness from a number of specific perspectives, we may in part replace evaluation and validation of a formal D-Case with computer-assisted verification of this property. However, verification of large and complex, formal D-Cases can itself become a complex task with no simple "yes" or "no" answers. Accordingly, this verification could conceivably be carried out using an assurance metacase—that is, an assurance case whose top claim is that the formal D-Case in question addresses openness from certain, specific perspectives.

For natural language-based assurance cases in GSN and CAE notations, proposals are beginning to emerge for meta-level arguments about object-level arguments, which in turn are about systems. Examples of these include the confidence argument, which claims that the arguments for system properties are highly reliable, and the metacase, which claims that the development process was appropriate in view of the record of systematic challenge and response to the (object-level) assurance case. These meta-level arguments are often constructed manually by examining the structure of the object level argument as can be revealed in graphical notations. In particular, the links between argument elements are examined to determine whether the content of linked elements is actually suitable for the link, whether there are insufficiencies, and so forth (e.g., "is this evidence sufficient for establishing this goal?").

In a formal D-Case, besides the linkage between argument elements, the content of each element also has rigorously formulated structure. Accordingly, the suitability and validity of its argument can be examined in greater detail and more rigorously. For example, suppose that a goal *Failure recovery time is within five minutes* (if a failure occurs now) is supported by a sole sub-goal *Failure recovery time as measured during testing on date XXXX was within three minutes* (which is evidenced by test results, etc.). The reasoning is insufficient as it lacks discussion of what could have changed since the time of testing (e.g., differences between the testing environment and the current operating environment, monitoring of the latter, configuration management of the running system, etc.). This inadequacy will not be detected by ordinary integrity checking if the writer of the case expressed this reasoning by a postulate. However, it would be detected if we are to verify that the argument has, for example, the property *Every goal with time-dependent meaning has at least one sub-goal with time-dependent meaning, or is supported by a monitor node.* Although the sense of "time-dependent" may vary from one argument to another, goals can be systematically classified based on their description in the formally specified vocabulary. The distinction between goals addressing properties of system models and goals addressing properties of the actual system may similarly benefit meta-level arguments concerning the validity of the case in question.

Use of patterns derived from best practice is one of the most effective means of ensuring the validity of a system's assurance case. With their structure and content, these patterns assist writers in covering all known important issues and presenting these issues clearly to the reader. The basic DEOS structure of Section 3.4 is one such example. These patterns also aid reviewers in identifying the nature of an argument as well as potential weak points of the argument due to its nature. An assurance metacase for a formal D-Case can be used to verify that the case under consideration has a specific structure and content, and also to verify checklist items derived from that structure and content. In many cases, patterns are not just instantiated but customized to suit the argument at hand. In those cases, the validity of the resulting argument cannot be justified by relying on the reputation of the base patterns. The assurance metacase can be used to verify that the customization retains the desirable properties of arguments that the original pattern was intended to bring about.

The two uses of assurance metacases in validation of formal D-Cases as described in the preceding paragraphs are complementary to each other. However, both require a meta-level vocabulary and the theory thereof in order to talk about the object-level vocabulary and theory and also about their significance in validation. A theory of validation needs to be developed, thereby addressing the concepts appearing in consensus building, accountability achievement, change accommodation, and failure response.

6.4 D-CASE IN AGDA—THE D-CASE INTEGRITY CHECKER

The Agda language is supported by the development environment and proof assistant of the same name. D-Case in Agda, an integrity checking tool for D-Cases, provides a link between the Agda interactive proof assistant and the graphical D-Case Editor. As such, it facilitates two-way conversion between Agda description of the reasoning part of a formal D-Case and argument trees in D-Case graphical notation. The style of Agda expressions for argument elements given in Section 6.2 is extended to include natural-language descriptions as text strings. On the D-Case Editor side, the internal data structure for argument elements is extended with a new attribute to store Agda expressions. The two-way conversion enables the combination of integrity checking of Agda descriptions using the proof-assistant functionality of Agda and reviews by domain experts using an intuitive display of argument trees (Fig. 6-4).

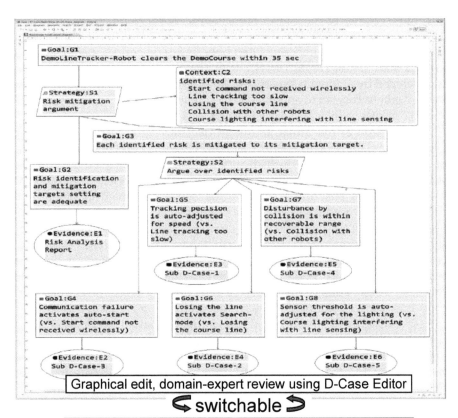

Graphical edit, domain-expert review using D-Case Editor

⤺ switchable ⤸

Verification, construction, generation using Agda

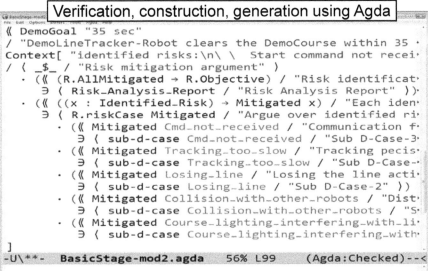

Fig. 6-4 Two Way Conversion.

Agda supports an interactive, refinement style development of code where an incomplete program containing holes ? can be type checked and where the holes ? can be refined and filled only in a type-correct manner. Let us assume that a D-Case argument tree with only natural-language descriptions is developed by a domain expert (see the upper half of Fig. 6-4). The process for an Agda technician to annotate this argument tree with Agda expressions and to complete a formal D-Case in Agda is as follows.

Conversion of the initial argument tree into Agda results in a formal D-Case whose theory part is empty and the reasoning part contains a skeleton of a <case> expression in which all argument elements are the holes ? paired with the original natural language descriptions. The technician analyzes the natural language description, adds to the theory part the declarations and definitions for vocabulary and axioms required for Agda description, and uses these to fill in the holes ? and complete the <case> expression. Type checking is performed at each stage so that the completed <case> expression is well-typed by construction. For example, in Fig. 6-4, the technician finds the term "identified risks" in the natural-language description of Context C2 and correspondingly adds to the theory part the declaration of the enumeration type `Identified_Risk`. Similarly, the predicate `Mitigated` is defined correspondingly to "... is mitigated to its mitigation target" in Goal G3, using data to be found in the evidence "Risk Analysis Report" in Evidence E1. This type and predicate are used to fill in the hole ? for G3, "Each identified risk is mitigated to its mitigation target", by the appropriate universal proposition, and the hole ? for Strategy S2 using the case-analysis function for the enumeration type `Identified_Risk`. At this time, Agda recognizes that there is only one way to consistently fill in the holes ? for G4 through G8, automatically generates the appropriate expressions, and fills in those holes ? Alternatively, the technician can fill in those holes ? using more readable expressions representing these same values (or types). In this case, Agda determines whether the expressions input by the technician have the expected values. In this way, integrity checking using the D-Case in Agda tool supports the *correct by construction* approach where formal D-Cases are interactively developed while maintaining integrity. The D-Case in Agda description can be converted back to an argument tree in D-Case graphical notation even if it is incomplete. In the tree form, the theory part of the D-Case in Agda is stored as attached data that is not usually displayed by D-Case Editor.

The D-Case in Agda tool is currently connected to D-Case Editor, but we plan to link the Agda proof assistant with D-ADD, so that D-ADD's dictionary functionality and the change management function can be enhanced with the formal D-Case approach.

6.5 EXAMPLE OF FORMAL ASSURANCE CASE IN AGDA

The following shows an example of how the assurance case argument from
Fig. 6-2 would be described as a formal assurance case using Agda.

```
{-# OPTIONS --allow-unsolved-metas #-}
module ExampleAssuranceCase where
open import Data.Sum
open import PoorMansControlledEnglish


--------------------------------------------------------------
-- Theory part
--------------------------------------------------------------
module Theory where
  postulate
    Probability-Type : Set
    impossible 1×10⁻³-per-year 1×10⁻⁶-per-year : Probability-Type
    _<_ : Probability-Type → Probability-Type → Set
  infix 1 _<_

  module C2-Control-System-Definition where
    postulate
      Control-System-Type : Set
      Control-System : Control-System-Type

  module C4-Hazards-identified-from-FHA where
    data Identified-Hazards : Set where
      H1 H2 H3 : Identified-Hazards
    postulate
      Probability-of-Hazard : Identified-Hazards → Probability-Type

  module C3-Tolerability-targets where
    open C4-Hazards-identified-from-FHA

    mitigation-target : Identified-Hazards → Probability-Type
    mitigation-target H1 = impossible
    mitigation-target H2 = 1×10⁻³-per-year
    mitigation-target H3 = 1×10⁻⁶-per-year

    Sufficiently-mitigated : Identified-Hazards → Set
    Sufficiently-mitigated h =
      Probability-of-Hazard h < mitigation-target of h

    postulate
      Eliminated : Identified-Hazards → Set
```

```
   argument-over-each-identified-hazard :
       H1 is Eliminated →
       H2 is Sufficiently-mitigated →
       H3 is Sufficiently-mitigated →
       ∀ h → h is Eliminated Or Sufficiently-mitigated
   argument-over-each-identified-hazard p1 p2 p3 H1 = inj₁ p1
   argument-over-each-identified-hazard p1 p2 p3 H2 = inj₂ p2
   argument-over-each-identified-hazard p1 p2 p3 H3 = inj₂ p3

 module C1-Operating-Role-and-Context where
   open C2-Control-System-Definition
   open C3-Tolerability-targets
   open C4-Hazards-identified-from-FHA

   postulate
     Software-has-been-developed-to-appropriate-SIL : Set
   postulate
     Acceptably-safe-to-operate : Control-System-Type → Set
   postulate
     argument-over-product-and-process-aspects :
       (For-all h of Identified-Hazards ,
        h is Eliminated Or Sufficiently-mitigated) →
       Software-has-been-developed-to-appropriate-SIL →
       Control-System is Acceptably-safe-to-operate

----------------------------------------------------------------
-- References to evidence
----------------------------------------------------------------
module Evidence where
  open Theory
  open C4-Hazards-identified-from-FHA
  open C3-Tolerability-targets

  postulate
    Formal-Verification   : H1 is Eliminated
    Fault-Tree-Analysis-2 : Probability-of-Hazard H2 < 1×10⁻³-per-year
    Fault-Tree-Analysis-3 : Probability-of-Hazard H3 < 1×10⁻⁶-per-year

----------------------------------------------------------------
-- Reasoning part
----------------------------------------------------------------
module Reasoning where
  open Theory
  open Evidence
  main =
    let open C1-Operating-Role-and-Context
        open C2-Control-System-Definition
```

```
in
Control-System is Acceptably-safe-to-operate
by argument-over-product-and-process-aspects
    • (let open C3-Tolerability-targets
            open C4-Hazards-identified-from-FHA
        in
        (For-all h of Identified-Hazards,
          h is Eliminated Or Sufficiently-mitigated)
        by argument-over-each-identified-hazard
            • (H1 is Eliminated
                by Formal-Verification)
            • (Probability-of-Hazard H2 < 1×10⁻³-per-year
                by Fault-Tree-Analysis-2)
            • (Probability-of-Hazard H3 < 1×10⁻⁶-per-year
                by Fault-Tree-Analysis-3))
    • (Software-has-been-developed-to-appropriate-SIL
        by {!!} )
{-# DCASE main root #-}
```

REFERENCES

[1] Kinoshita, Y. and M. Takeyama. 2013. Assurance Case as a Proof in a Theory: towards Formulation of Rebuttals, in Assuring the Safety of Systems—Proceedings of the Twenty-first Safety-critical Systems Symposium, Bristol, UK, 5–7 February 2013; C. Dale and T. Anderson (eds.). SCSC, pp. 205–230, Feb. 2013.

[2] Takeyama, M. 2011. D-Case in Agda' Verification Tool (D-Case/Agda). http://wiki.portal. chalmers.se/agda/pmwiki.php?n=D-Case-Agda.D-Case-Agda

[3] Takeyama, M., H. Kido and Y. Kinoshita. 2012. Using a proof assistant to construct assurance cases, Fast Abstract in Proceedings of Dependable Systems and Networks (DSN).

[4] The GSN Working Group. 2011. GSN Community Standard, Version 1.

[5] Kelly, T. and R. Weaver. 2004. The Goal Structuring Notation—A Safety Argument Notation, in Proceedings of the Dependable Systems and Networks 2004 Workshop on Assurance Cases.

[6] Bishop, P. and R. Bloomfield. 1998. A Methodology for Safety Case Development. Industrial Perspectives of Safety-Critical Systems: Proceedings of the Sixth Safety-critical Systems Symposium. Birmingham.

[7] Toulmin, S. 2003. The Uses of Argument, Cambridge University Press, 1958. Updated edition published in 2003 by the same publisher.

[8] Basir, N. et al. 2009. Deriving Safety Cases from Machine-Generated Proofs. In Proceedings of Workshop on Proof-Carrying Code and Software Certification (PCC'09), http://eprints. soton.ac.uk/id/eprint/271267, Los Angeles, USA.

[9] Hall, J., D. Mannering and L. Rapanotti. 2007. Arguing safety with problem oriented software engineering. In the Proceedings of High Assurance Systems Engineering Symposium, HASE'07, 10th IEEE, pp. 23–32.

[10] Agda Team, Agda Wiki. http://wiki.portal.chalmers.se/agda/pmwiki.php. Accessed 5 October 2012.

[11] Tokoro, M. (ed.). 2012. Open Systems Dependability—Dependability Engineering for Ever-Changing Systems, CRC Press.

7

D-RE—The DEOS Runtime Environment

The DEOS Runtime Environment (D-RE) provides the execution environment—or in general terms, the operating system—for application programs in the DEOS Architecture in order that the DEOS Process may be carried out. In this chapter, we first look at its design concept and basic structure, and then at a web system application and robots as case studies. Our approach to security and methods developed accordingly for D-RE are also described.

7.1 DESIGN CONCEPT AND BASIC STRUCTURE

The purpose of D-RE is to provide services to operating systems and application programs that run on it. You can find D-RE at the bottom right of the DEOS Architecture illustrated in Fig. 3-3 in Chapter 3. Figure 7-1 shows one embodiment of this environment, D-RE(4), which will be explained later, as an example. Multiple operating systems and applications must be able to execute within reconfigurable, independent runtime environments in accordance with the stakeholders' dependability requirements, and for this reason, we have defined five critical functions and four optional functions for D-RE. We introduce and describe these functions in detail in Subsection 7.1.1. The critical functions are then defined as five abstracted components that makeup D-RE, and these are described in Subsection 7.1.2. A few D-RE options for different application areas are then presented in Subsection 7.1.3.

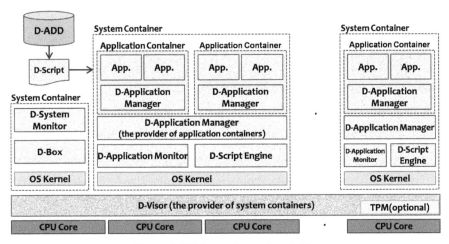

Fig. 7-1 D-RE(4) as an Example of D-RE.

7.1.1 D-RE Functions

After applying what-if analysis to case studies of past system failures (see Appendix A.3 for details), we defined the following five functions as being critical requirements for D-RE:

- Monitoring
- Reconfiguration
- Scripting
- Secure recording
- Security.

In addition, we have also defined the following four optional functions:

- Resource quota
- Undo
- Migration
- System status recording.

(1) Critical functions

Each of the five critical D-RE functions—monitoring, reconfiguration, scripting, secure recording, and security—are now described in detail.

A. Monitoring

D-RE's monitoring function monitors the condition of applications, operating systems, and other components of the information system and detects any deviation thereof from the Ordinary Operation state. D-RE provides the following sub-systems for this purpose:

- D-Application Monitor

  The purpose of a D-Application Monitor is to monitor the condition of applications and to collect logs as specified by D-Case monitor nodes. It provides two distinct mechanisms—namely, monitoring of resource abuse and tracing of application-specific events. These mechanisms make it possible for deviation of the operational information system from the *Ordinary Operation state* to be detected.

- D-System Monitor

  The primary role of a D-System Monitor is to detect key loggers, rootkits, and other kernel intrusion, which it does using functionality capable of detecting kernel-level anomalies.

B. Reconfiguration

The purpose of the D-RE reconfiguration function is to change the internal structure of the information system. It would likely be applied upon the detection of an anomaly or in the event of failure. Reconfiguration requires functionality for isolating operating systems from one another or applications from one another within specific logical partitions (i.e., units having a suitable resource-allocation policy). D-RE provides two types of container as isolation units.

- System containers

  A system container provides logical partitions for isolating system services from one another.

- Application containers

  An application container provides logical partitions for isolating applications from one another.

Each system container allows an operating system to run independently of others. Meanwhile, each application container allows applications to run independently of others. Each container includes a runtime environment for its own programs, and restarts are performed on an individual-container basis. The necessary container attributes are listed in Table 7-1, and those to be implemented depend on the stakeholders' dependability requirements. Meanwhile, core system services including clocks and measurement units must be accessed via D-RE and isolated from the system and application containers.

Table 7-1 Container Attributes.

Attribute	Description
Address space	Independence of address space referenced by CPU
Namespace	Independence of space comprising file names or OS IDs such as process IDs
Physical memory	Independence of physical memory space referenced by CPU
Cache memory	Independence of cache memory area
CPU scheduling	Assurance probability for maximum CPU utilization
CPU allocation	Independence concerning allocation of CPU cores to individual virtual machines
I/O bandwidth	Assurance probability for maximum I/O bandwidth
Bus bandwidth	Assurance probability for maximum bus bandwidth
Interrupt	Routing probability for interrupt to suitable virtual machine
Time-of-day	Control probability for each virtual machine's independent actual time
Privilege	Independence between different privilege levels

C. Scripting

In the event that the information system has deviated from the *Ordinary Operation state* or shows signs of doing so, D-RE provides a script runtime environment so that scripts can restore or maintain that state. The role of a script is, therefore, to function as a program concerned with non-functional requirements, running separately from the application program in order to respond to failures based on service continuity scenarios. Such a script is called D-Script, and D-Scripts are executed either automatically or manually by operators.

D-Scripts are required to reduce the load associated with maintaining system dependability primarily due to operators, and they would typically (1) collect operation-related information and (2) perform system reconfiguration. D-Scripts are also subject to stakeholders' agreement, and a D-Script pattern has been defined to make them easier to comprehend. The pattern comprises two items—*an anomalous system state* and *an action for restoring the normal state*—and these are incorporated into the D-Case description. We have also defined D-Script tags in the interest of striking a good balance between human readability and efficiency in machine-based processing. In order, meanwhile, to support the use of D-Script execution results as evidence, restrictions on D-Scripts are made so that they provide at least *recording of success or failure* and *description of recovery actions upon failure*.

D-Script Engine is the D-RE component that allows D-Scripts to run therein. As such, it ensures that these scripts can be executed in a trusted environment within the information system. D-Script Engine gets D-Scripts from D-ADD and executes them. Results of script execution are stored in a D-Box (described below), and operators are notified upon the development of any condition that would prevent D-Scripts from being executed. D-Script and D-Script Engine are both described in more detail in Chapter 8.

D. Secure recording

D-RE must safely and reliably record logs detailing past and pre-failure system conditions as well as D-Script execution logs. It has, therefore, been provided with D-Box log recording areas. D-Box records must be tamper resistant so that D-RE can present service records, logs, and other recorded data to stakeholders as authentic information. For this reason, each D-Box is provided with functionality for: (1) interfacing with D-Application Monitor and D-System Monitor; (2) recording of logs based on stakeholders' agreements (concerning, for example, events, anomalies, and software updates); and (3) managing access authentication and authorization. Meanwhile, this D-RE component also works with D-ADD (described in Chapter 9) to ensure consistency of system state records, particularly those concerning dependability.

E. Security

D-RE must ensure that the above-described critical functions are executed in a secure manner, and it has been provided with a number of features, such as access control, authentication and authorization management, and system takeover prevention. It is recommended that these features be configured from a Trusted Computing Base (TCB) [1] comprising hardware, software, and procedural components so that a security policy can be enforced. The TCB portion of D-RE will be the only part of the information system that cannot be modified after service startup, and as such, it must be properly configured upon system deployment. While D-RE features a range of security mechanisms, particular attention is paid to secure running of the operating systems themselves. This ensures that the other security mechanisms can be configured on a secure operating system, and D-Visor and D-System Monitor help to make this possible. In brief, D-Visor isolates the operating systems from one another, and D-System Monitor can detect deviation of operating systems from their normal state. These D-RE security mechanisms are described in more detail in Section 7.4.

(2) Optional functions

In addition to the five critical functions described above, D-RE can be provided with four optional functions in order to enhance its usefulness. Each of these functions is described below.

A. Resource quota

With resource quota, D-RE containers can limit the amount of CPU and memory resources (i.e., CPU utilization and memory utilization, respectively) used by the programs they contain in order to prevent any other containers from being adversely affected. If a program in one specific container were to run out of control and try to monopolize the CPU, this function would ensure that sufficient CPU resources are still available to programs in the other containers; similarly, it could prevent depletion of system memory resources due to over use by one container. D-RE resource quota can be dynamically configured after container startup; accordingly, resources could be limited based on the results of system or application monitoring so as to prevent failure and ensure continuity of services, albeit with some delay.

B. Undo

It is often the case that new services are added to existing systems or a corrective code is introduced to fix a problem. However, no amount of repetitive testing can guarantee in advance that this type of additional or corrective code will run exactly as intended. In the worst case scenario, meanwhile, it could knock the entire system down and interrupt the services currently being provided to users. If it were possible to restore the condition before introducing the new code in such a situation, the original services could, at the very least, be provided. For this reason, support for cancellation of such action—or undo—is highly important.

Generally speaking, it is not easy to implement an undo feature whereby a sequence of processes can be cancelled in reverse order. A common alternative solution is to store the state of the system at the point in time before the start of each process and to then return to these stored states when required. D-RE provides functionality for storing the state of each system and application container at any time and starting processing again from any such state.

Storing of the state of a system can be done with the container either completely or temporarily stopped. In the case of the former, execution of all of the programs in the container will be terminated, and the restart procedure will be identical to a normal program restart. In the latter case, the execution state of the container is stored while it is temporarily stopped simply by suspending the allocation of CPU resources, meaning that the container's programs remain in operation and the corresponding memory content is also preserved. Accordingly, this memory content can be used to implement the restart, and the temporarily halted programs simply continue their processing. In terms of D-RE, the former functionality is referred to as *the saving and reloading of snapshots*; the latter is referred to as *checkpoint and restart*.

C. Migration

D-RE supports the transfer of guest operating systems from one system container to another—a capability commonly known as virtual machine (VM) migration. In addition to migration within a specific host machine, container-unit migration between physically different machines connected via a network is also supported. In order to facilitate the latter type of migration, the destination machine must support the use of system containers. In addition to migration of a system container's guest operating system after first temporarily halting it, D-RE also supports live migration—in other words, relocation of a guest operating system as it continues to run. While this functionality can certainly help in the avoidance of failure caused by hardware issues, it is more useful in terms of service continuity when hardware has to be replaced or repairs made.

D. System status recording

In addition to application logs, records of what was occurring inside a system immediately before a failure can be extremely useful in identifying the cause. However, failure can conceivably occur at any time, and therefore, information describing the prior state of the system can only be made available by constantly recording in much the same way as an aircraft flight recorder or the event data recorder found in some automobiles. System status recording is the D-RE mechanism whereby this is achieved.

7.1.2 D-RE Components

In order to achieve the functionality described above, we have defined the following six D-RE components: (1) D-Visor, (2) D-System Monitor, (3) D-Application Manager, (4) D-Application Monitor, (5) D-Box, and (6) D-Script Engine. This section provides implementation guidelines for each.

(1) D-Visor

D-Visor is a virtual machine monitor used to implement the system containers that secure the independence of operating system components. Each system container takes the form of a VM on each of which an independent guest operating system runs. Failures or anomalies occurring in one specific system containers are prevented from propagating to others.

D-Visors for D-RE are currently implemented using Linux KVM [2], ART-Linux [3, 4], SPUMONE [5], and D-Visor86 [6]; accordingly, the best solution can be determined based on the system's dependability requirements. With Linux KVM, for example, system containers would be implemented using Linux kernel-based VMs. This allows for the realization of system containers satisfying the requirements of the address space, namespace, CPU scheduling, CPU allocation, interrupt, time-of-day, and privilege attributes identified in Table 7-1.

Meanwhile, D-Visor86 has been developed at the University of Tsukuba as a virtual machine monitor for multicore processors, and it allows a modified Linux kernel to run on each processor core of x86 multi-core CPUs (including hyperthreaded CPUs). Using this approach, a D-System Monitor running on Linux on one processor core can monitor Linux on the others.

D-Application Manager APIs and commands use Linux KVM to implement D-Visors. With an independent guest operating system running in each system container, the startup, shutdown, and other similar operations of the guest operating system in one does not affect the others. And because running of different guest operating systems in each system container is possible, independent applications can be housed in different containers, thereby preventing a failure in any one affecting the rest of the system.

(2) D-System Monitor

In cooperation with D-Visor, D-System Monitor provides functionality for monitoring system containers. It monitors VMs from the outside without affecting these operating systems in any way and can observe guest operating systems running in the containers for anomalous behavior due to tampering and the like (Fig. 7-2).

The D-Visor provides the D-System Monitor with functionality for monitoring the I/O requests of the individual VMs and also for directly referencing the memory of the operating systems being monitored. Using data acquired in this way, the monitoring mechanism of the D-System Monitor detects anomalous behavior in the operating system being monitored. At

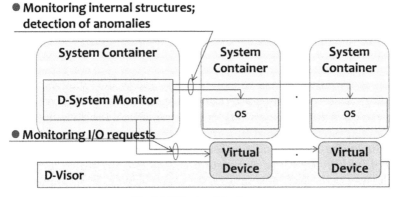

Fig. 7-2 D-System Monitor.

present, the following three monitoring mechanisms developed at Waseda University and Keio University can be utilized for this purpose:

- *FoxyKBD* [7] observes operating system behavior upon keyboard input. It artificially generates a large volume of keyboard input and detects malware attempting to intercept this based on the resulting I/O requests.

- *RootkitLibra* [8] monitors for file concealment. Working with files from NFS-mounted directories, it compares information from files visible to the system and files in NFS packets in order to detect concealment and file-size tampering.

- *Waseda LMS* [9] is a mechanism that monitors for process concealment. Directly referencing the memory of the operating system being monitored, it compares kernel process-list and run-queue data.

Each of these monitoring mechanisms can be used by the D-System Monitor running on D-Visor86 as developed at the University of Tsukuba. A QEMU/KVM version incorporating D-System Monitor has also been developed, allowing the above three types of monitoring mechanism to be used with KVM system containers.

(3) D-Application Manager

D-Application Manager provides application containers as a means of isolating applications from one another as well as a library of functions that support both the output of application logs and shutdown processing. These application containers are not VMs but lightweight containers on an operating system. D-Application Manager uses Linux Containers (LXC) [10] as application containers controlled via APIs and commands. Multiple application containers run on a specific operating system, and an operating-system failure can thus affect all of them; however, processes within each application container are isolated from those of the other containers and not affected by them. Application containers allow multiple processes to be treated as a single group, meaning that CPU and memory resources can be restricted on a group-specific basis. By placing applications running on the same operating system in different application containers, the corresponding CPU and memory resources can be isolated from one another, thereby mitigating the effect of failure of one application on the others. Application containers have also been provided with snapshot save and load functionality. However, LXC checkpoint and restart functionality has not yet been implemented, so these functions are not available for D-RE application containers.

For both system and application containers, the utilization sequence comprises *create*, *start*, *stop*, and *destroy* steps in that order. These steps can be executed using commands such as *dre-sys* and *dre-app* and libraries such as *libdresys* and *libdreapp*. In addition, commands and libraries also provide access to checkpoint and restart as well as snapshot save and load functionality.

D-Application Manager provides functionality supporting log output and shutdown processing in the form of a *libdaware* library, which is used when creating applications. It also includes functions and macro definitions that make it easy to specify the format for log output to syslog. Sending a signal providing advance notification of termination can make it possible to evacuate critical data before an application is terminated and to store necessary data upon restart. D-Application Manager functions support the implementation of this type of termination process. Because process IDs used within specific application containers differ from those visible from the outside, it may be necessary to identify a certain process ID in order to send a termination notification signal from outside the container in question. For such a case, D-Application Manager provides a mapping system (/proc/daware) for internal and external process IDs.

(4) D-Application Monitor

A D-Application Monitor examines the inter-relationships between multiple events received from its D-Application Manager, applies event correlation to identify phenomena (potentially anomalous) suggested by these events in their entirety, and performs the necessary action in order to deal with these phenomena.

Figure 7-3 provides a block diagram illustrating the structure of D-Application Monitor (including some components yet to be implemented). D-Application Monitor is configured with a built-in web server and is implemented as a process accessed using representational state transfer (REST)

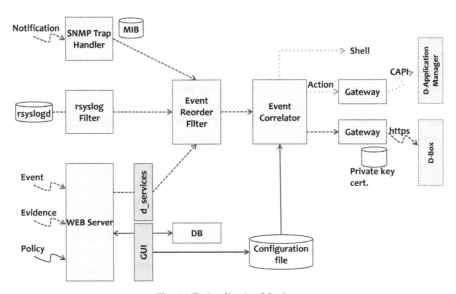

Fig. 7-3 D-Application Monitor.

APIs. In consideration of recent trends towards implementing and using a number of simple web servers, we have added support for REST APIs to make D-Application Monitor more convenient.

A D-Application Monitor receives events from its D-Application Manager. If necessary, it may be extended going forward to accept SNMP notifications (traps) and *rsyslogd* messages. D-Application Monitor sends all received events—whether from web servers, SNMP notification, or *rsyslogd*—to the Event Reorder Filter, which compensates for source-specific delay by correcting event times within a certain window and then reordering them. Once arranged in chronological order in this way, the events are sent to the Event Correlator in order to have the inter-relationships between them examined. The Event Correlator is implemented with the open-source Simple Event Correlator (SEC), which uses finite state machines to receive events and realize state transition.

For example, SEC can realize the state transition and the corresponding actions shown in Fig. 7-4. Actions can take the form of shell script execution, sending of signals to other processes, and so forth. SEC runs on the basis of a configuration file describing finite state machines and actions pursuant to state transitions. D-Application Monitor allows this configuration file to be uploaded and edited via a web server.

D-Application Monitor also provides functionality for recording in a D-Box the state of the system immediately before the occurrence of a failure. A ring buffer is used to record constantly the actions of the operating system and applications over a certain period of time (e.g., 5 minutes). Thus, it becomes possible in the event of a failure to determine exactly what was occurring immediately before it, which can be highly useful when trying to pinpoint the cause. Certain VMs feature record and replay functions, which can be used not only to record the actions of the entire system, but also to reproduce the failure condition. If, however, it is sufficient to simply record actions without the need to reproduce them, *ltrace* and *strace* commands on Linux could be used as an alternative solution. These commands can be conveniently used to record library and system calls, which will provide valuable clues when analyzing the behavior of applications. However, these commands are not intended for regular use, and doing so can lead to the generation of massive output files; similarly, if used with multiprocess applications, *ltrace* or *strace* for just one of

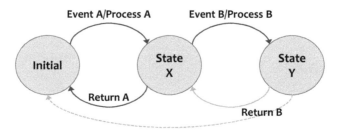

Fig. 7-4 State Transition in Simple Event Correlator.

the processes can cause an overall processing bottleneck. We modified these commands for our implementation of state recording functionality in order to avoid the problems described.

(5) D-Box

Using a public key infrastructure (PKI), the D-Box provides functionality for storing of log records with additional information that makes it possible to detect data falsification. As shown in Fig. 7-5, the D-Box of a system resides at the bottom of a hierarchical structure based on public key certificates issued by a root certificate authority (CA), meaning that the validity of its own public key can be assured.

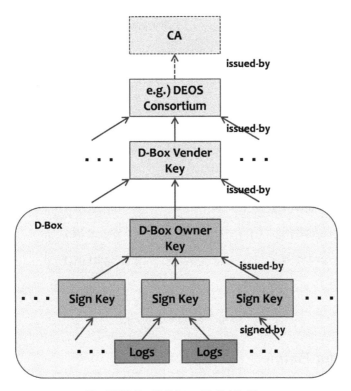

Fig. 7-5 D-Box Validity with Public Key.

When the D-Box is initialized, the manufacturer installs a unique D-Box Owner key (in the form of PKCS#12) in it. The key is signed with the manufacturer's own private key and the manufacturer's public key is, for example, signed with the DEOS Center private key in a sequence that ultimately leads back to a trusted root certificate authority. Thereafter, the

D-Box generates pairs of public and private RSA keys for signing log records during operation. These key pairs are regenerated as appropriate in order to prevent the same private key from being used over an extended period of time.

When the D-Box receives log records via HTTPS (TLS/SSL), it interprets them as octet strings and produces MD5 hashes (Digests) with the newest signing private key it contains (Fig. 7-6). When a program acquires log records from the D-Box, it also retrieves the corresponding encrypted hashes and decrypts them using the signing public key for that D-Box. The program then creates MD5 hashes from the acquired log records and compares them with the decrypted hashes in order to detect log-record falsification (Fig. 7-7). Figure 7-8 shows the configuration of D-Box. Similarly to D-Application Monitor, a D-Box also provides REST APIs.

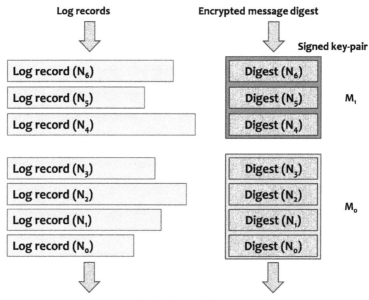

Fig. 7-6 Log and Digest.

(6) D-Script Engine

D-Script Engine is tasked with the safe and reliable execution of D-Scripts, and it has the same execution permission as the corresponding program. D-Scripts control applications using APIs provided by D-RE, and therefore, D-Script operations are limited by the specific implementation of the D-RE in which D-Script Engine is installed. The following functions are, however, defined as essential: (1) storing of D-Script execution results in a D-Box, and (2) notification of operators when D-Scripts cannot be used.

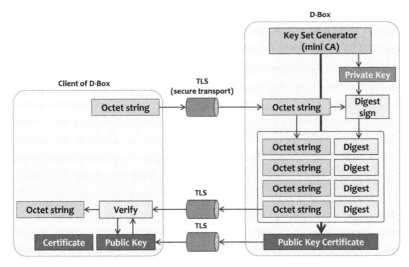

Fig. 7-7 D-Box and Usage Thereof.

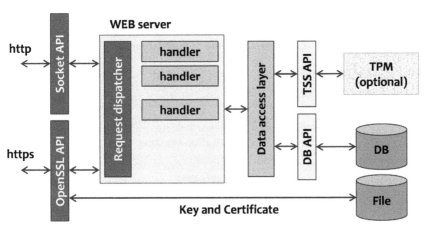

Fig. 7-8 D-Box Configuration.

In D-RE, it is D-ADD that supplies the D-Scripts to D-Script Engine. Meanwhile, D-Script Engine features functionality for ensuring that only D-Scripts agreed upon by the stakeholders can be used during operation. A more detailed description of this D-RE component is provided in Chapter 8.

7.1.3 Tailoring of D-RE to Information Systems

As an implementation architecture, D-RE must be tailored to the information system with which it will be used on the basis of that system's dependability requirements as well as its functionality. In this section, we look at four D-RE implementation case studies. In each, the system in question will be required to satisfy a different set of dependability requirements. The corresponding runtime environments are referred to as D-RE(1) through D-RE(4):

- D-RE(1): Configuration for a simple application
- D-RE(2): Configuration for a multicore embedded system
- D-RE(3): Configuration for a real-time application
- D-RE(4): Full-scale configuration

The first of these, D-RE(1), is shown in Fig. 7-9. This example does not contain any D-Visors or D-System Monitors; accordingly, the operating system could constitute a dependability weak spot. The D-Box must be specially configured using the functionality of the underlying operating system. D-Application Manager, D-Script Engine, and D-Application Monitor run under the protection of the underlying operating system, and therefore, the dependability of execution of the applications depends on that of the operating system.

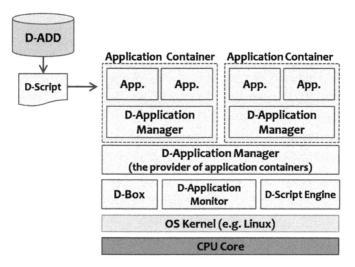

Fig. 7-9 D-RE(1): Configuration for a Simple Application.

The configuration of D-RE(2) is shown in Fig. 7-10. Here, the DEOS Runtime Environment has been tailored for an embedded system featuring multicore processors, and it uses the SPUMONE virtualization layer [5] optimized with D-Visor and D-System Monitor. In this example, no special equipment is required in order to virtualize the underlying hardware.

As shown in Fig. 7-11, D-RE(3) is configured for the execution of real-time applications for robots and the like. This D-RE uses the ART-Linux visualization layer optimized for D-Visor, D-System Monitor, and the underlying operating

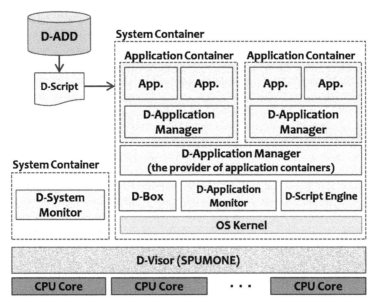

Fig. 7-10 D-RE(2): Configuration for a Multicore Embedded System.

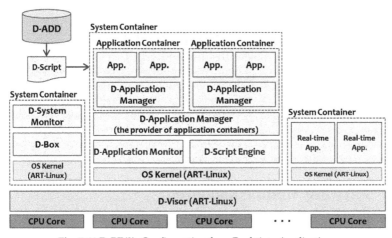

Fig. 7-11 D-RE(3): Configuration for a Real-time Application.

system. As a Linux kernel extension, ART-Linux has binary compatibility for Linux applications. More details are provided in Section 7.3.

As a full configuration of the DEOS Runtime Environment, D-RE(4) is shown in Fig. 7-1. The system container on the left-hand side of this figure is provided for the D-Visor, D-System Monitor, and D-Box. Depending on security requirements, however, it may be necessary to allocate these components to different system containers individually. Each of the other system containers is for the applications' operating systems, D-Script Engines, D-Application Monitors, and D-Application Managers. Each D-Application Manager provides application containers for application programs. The D-Application Manager in each of the application containers is also used as a proxy for access to D-RE components such as D-Application Monitor and D-Script Engine.

REFERENCES

[1]   DoD 5200.28-STD Trusted Computing System Evaluation Criteria (Orange Book), December 26, 1985.
[2]   http://www.linux-kvm.org/
[3]   Kagami, S., Y. Ishiwata, K. Nishiwaki, S. Kajita, F. Kanehiro, W.N. Yoon, N. Ando, Y. Sasaki, S. Thompson and T. Matsui. 2012. Design and Implementation of ART-Linux Capable of Using Multiple CPU Cores by Combining AMP & SMP in Proceedings of 17th Robotics Symposia, pp. 521–526, Hagihonjin, Hagi City, Yamaguchi Prefecture, March 2012 (in Japanese).
[4]   http://www.dh.aist.go.jp/jp/research/assist/ART-Linux/
[5]   Nakajima, T., Y. Kinebuchi, H. Shimada, A. Courbot and T.-H. Lin. 2011. Temporal and Spatial Isolation in a Virtualization Layer for Multi-core Processor based Information Appliances, 16th Asia and South Pacific Design Automation Conference, pp. 645–652.
[6]   http://www.jst.go.jp/crest/crest-os/tech/D-SystemMonitor/index-e.html
[7]   Kono, K. VMM-based Approach to Detecting Stealthy Keyloggers, http://www-archive. xenproject.org/files/xensummit_tokyo/21_KenjiKono-en.pdf
[8]   Kono, K., P. Rajkarnikar, H. Yamada and M. Shimamura. VMM-based Detection of Rootkits that Modify File Metadata, Research Report, Computer Architecture Study Group, Information Processing Society of Japan (in Japanese).
[9]   Shimada, H., A. Courbot, Y. Kinebuchi and T. Nakajima. 2010. A Lightweight Monitoring Service for Multi-Core Embedded Systems. In Proceedings of the 13th Symposium on Object-Oriented Real-Time Distributed Computing, pp. 202–210. doi: 10.1109/ISORC.2010.12.
[10] http://lxc.sourceforge.net/

7.2 WEB SYSTEM CASE STUDY

7.2.1 Software Module Makeup

In this section, we study D-RE implementation in a web-based prototype to simulate a CD online shopping system using Apache, Tomcat, and MySQL (Fig. 7-12). In this environment, Apache, Tomcat, and MySQL run in separate, dedicated system containers. In addition, the application *Nagios* (http://www. nagios.org/) is used to monitor the computer system and the network. The *ab* tool for benchmarking Apache HTTP servers is employed in order to simulate multiple simultaneous requests from multiple browsers on the network.

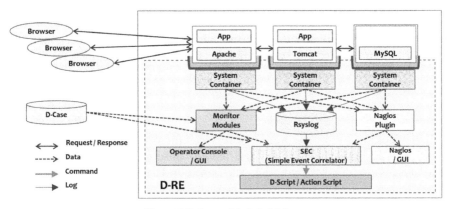

Fig. 7-12 Software Module Makeup.

Service scenarios

This information system has a classic configuration comprising web, application, and database servers. In addition, these servers are constantly monitored via an operator console. In terms of software configuration, each of the servers is implemented on a different system container within D-RE. The operator console consolidates logs from the individual system containers using *rsyslog*, and the servers' resources and performance are monitored by means of monitor modules and *Nagios* plugins. If a monitor module detects a failure, it simultaneously notifies the operator via the GUI and passes the relevant information to SEC [1]. SEC then executes D-Scripts based on this information. The following scenarios illustrate how D-Case and D-RE deal with different types of system failure.

Scenario 1

As a result of the addition of a new service, traffic increases beyond the design level. In line with the corresponding D-Script in the D-Case, the system executes an undo with respect to the service addition and restores the normal state.

Scenario 2

Server response times slow down abruptly. In line with the corresponding D-Script in the D-Case, the system terminates unsuitable batch jobs and restores the normal state.

Scenario 3

Memory usage exceeds the quota for the system. In line with the corresponding D-Script in the D-Case, the system restarts the server and deploys diagnostic modules to identify the cause.

Scenario 4

When checking the following day's services for potential issues as part of regular inspection, it is found that the delivery of services will not be possible. The cause is identified as expiration of a license, and the license is thus renewed.

7.2.2 System Software Configuration

Figure 7-13 shows the software configuration of a system using D-RE. The monitoring sub-system is implemented using monitor module instances corresponding to the D-Case monitor nodes. Meanwhile, the analysis sub-system is implemented by sending events detected by these monitor module instances to finite state machines realized with SEC and then performing event analysis on the basis of the inter-relationships between multiple events.

7.2.3 DEOS Programs—from Development to Execution

The DEOS dependability functions are developed separately but in parallel with the development of the application programs for this CD online shopping system, and they are executed as shown in the following sequence:

(1) Argue and agree on service requirement specifications (thereby defining the D-Cases)
- Identify corporate management policy and business model
- Select and agree on a business continuity plan (BCP)
- Select and agree on SLA parameters
- Create SLA-related monitor and action nodes

(2) Select monitoring targets and parameters for the monitor nodes
- Reuse existing monitor modules where possible
- Define the required specifications for new monitor modules
- Prepare external specifications of the D-Script description files for monitoring

(3) Define detailed requirements for analysis nodes
- Define analyzers using finite state machines
- Prepare external specification of the D-Script description files for analysis using the SEC configuration file template

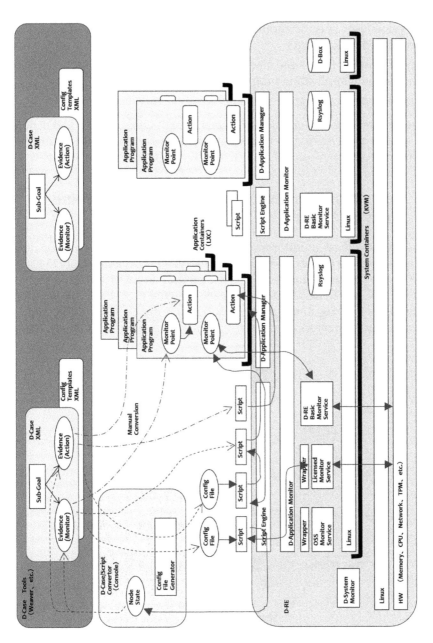

Fig. 7-13 System Software Configuration Using D-RE.

(4) Develop application programs—i.e., the CD online shopping system in this example

(5) Develop user-specific monitor modules in parallel with application program development

(6) Implement the finite state machines and develop user-specific actions in parallel with application program development

(7) Run the application programs on D-RE

The last steps initiate creation of instances of D-Case monitor and action nodes so that the DEOS dependability functions run in parallel with the application programs.

7.2.4 Mechanism for Executing D-Case Monitor and Analysis Nodes

Figure 7-14 shows how monitor module instances and SEC configuration files are created from the D-Case using D-Case Weaver[1] [2]. As a first step, D-Case Weaver creates an entry for each monitor module in a D_Case_Node table within the D-Case. To do so, D-Case Weaver has the user select the monitor module corresponding to the monitor node from the D_Script table, and it adds the set of parameters required for creation of an instance thereof to the

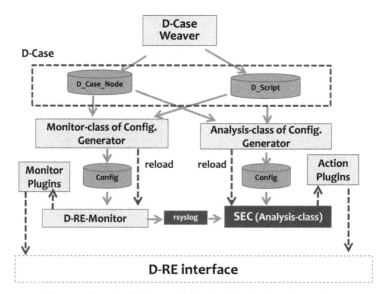

Fig. 7-14 Creating Monitor and Analysis Nodes from the D-Case.

[1]For more details, refer to *DEOS-FY2013-CW-01J: D-Case Weaver Specification, Introduction, Users' Guide.*

monitor node. In other words, D-Case Weaver associates D-Case monitor nodes with D-Script monitor modules. Following this, it calls the configuration file generator for monitoring and the configuration file generator for analyzers.

The configuration file generator for monitoring joins the D_Case_Node table and the D_Script table, identifies the monitor module corresponding to each monitor node, applies the parameter set from the monitor node to its monitor module template, and creates a monitor module instance. Similarly, the configuration file generator for analyzers joins monitor nodes and D-Scripts in the SEC configuration file templates in order to apply specific parameter values from the monitor nodes to the SEC configuration file template and create the corresponding SEC configuration files. The following sections describe how monitor and analysis nodes are implemented.

Creating monitor modules

Each monitor takes the form of a monitor module instance customized using a specific set of parameters. The D-RE monitor modules for this system are written in Python and implemented as sub-classes of the PluginBase class. Accordingly, the term "monitor instance" refers to an instance of that sub-class—an instance created by allocating a specific parameter set to the constructor of the sub-class representing monitor modules.

Figure 7-15 shows a typical example of how a specific monitor instance is created from a monitor module, which is defined as a D-Script including a set of parameters that can be overwritten, and specific values for those parameters as described in a monitor node in the D-Case.

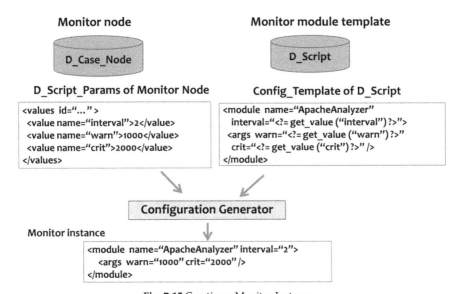

Fig. 7-15 Creating a Monitor Instance.

Creating analyzer modules

In implementing finite state machines using SEC, the individual machines and actions are created, in the same way as monitor modules, by applying specific sets of parameters to corresponding finite state machine and action templates.

Module templates for monitor and analyzer

Module templates take the form of the XML files containing the following:

- An identifier for a monitor module or finite state machine template
- A functional description, the author's name, a set of configurable parameters, and descriptions thereof
- A template with placeholders for parameter values and used to generate instances of monitors, finite state machines, or actions

7.2.5 Creating D-Case Monitor Nodes

D-Case evidence nodes related to monitoring describe how programs running on D-RE are to be monitored. Normally, the evidence nodes required for monitoring such that the service requirement specifications may be met are identified when creating the D-Case during program design.

Each monitor node identifies a monitor module and specifies values for the set of parameters required to run an instance of that module.

7.2.6 Creating D-Case Analysis Nodes

Instances of monitor modules observe the state of a specific part of the system in order to detect failure, anomalies, signs of potential failure, and so on. Generally speaking, multiple instances monitor for the same specific failure from their own particular viewpoint and report individually. In situations where individual monitor module instances report on multiple types of failure, these reports can narrow down the possible causes. Accordingly, analyzers must be implemented so as to interpret failure reports (i.e., events) from multiple instances of monitor modules in an all-inclusive manner and thereby identify the most likely cause.

This all-inclusive interpretation of multiple events could conceivably be achieved by applying conditional branching (using *if* statements) to event detection and then combining these conditionals; however, this approach has a serious drawback in that the number of conditional sequences increases exponentially when the order of execution of the conditional branches must be rearranged to cover all possible combinations of occurrence timing for multiple events. In addition, combinations of conditional statements are not well suited to the detection of multiple events occurring within a specific (short) period of time. As described below, finite state machines are extended so that we can simplify the process of interpreting multiple events.

Figure 7-16 shows a finite state machine that performs an action A and transitions to a state X when an event A occurs while in its initial state, and that performs an action B and transitions to a state Y when an event B occurs while in state X.

On the other hand, Fig. 7-17 shows a finite state machine that transitions to state Z when event A and event B occur in any order.

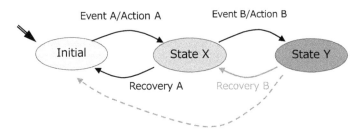

Fig. 7-16 State Transitions and Actions.

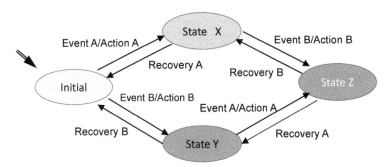

Fig. 7-17 Processing of Events Irrespective of Occurrence Sequence.

Meanwhile, specifying the maximum duration for which individual states can be maintained facilitates the detection of multiple events occurring within the same period of time. For example, we could set the maximum duration of state X to 30 seconds, meaning that if event B were to occur within 30 seconds of event A, the finite state machine would transition to state Z and execute a specific process.

In this type of analyzer implementation using finite state machines, each machine transitions to the state corresponding to a sequence of related events that it receives, thereby making it easier to identify failures with multiple events as symptoms. Furthermore, multiple instances of finite state machines corresponding to specific D-Case nodes could be run in parallel in order to simultaneously analyze mutually independent sequences of events in terms of the entire D-Case. We could also send the same event sequence to multiple finite state machines to have it analyzed simultaneously from a range of different viewpoints. This type of approach to the analysis of event sequences is clearly more convenient than using combinations of conditional statements.

Certain events can happen sporadically even during normal operation, such as when network-related failure occurs. In Fig. 7-18, we can see an example of how a succession of these events could be handled. In response to the first such event, the finite state machine writes a log entry in order to maintain records and then transitions to a *Burst* state, wherein it can remain until a timeout duration has elapsed. If subsequent events were to occur while in that state, it would remain inactive until the nth such event, which it would interpret as evidence of a permanent failure. The finite state machine would then take the prescribed action, such as notifying the operator, before transitioning to a *Dormant* state. In this state, the finite state machine ignores all similar events. In specific terms, SEC can be used to create the required configuration files based on D-Case action nodes.

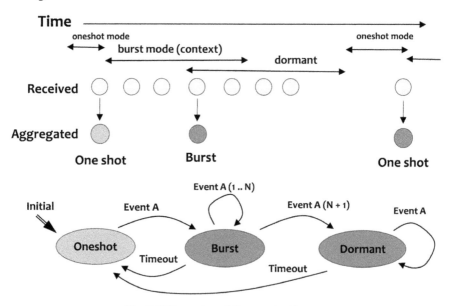

Fig. 7-18 Processing of Consecutive Events.

7.2.7 Examples of Action Commands

In this case study, the system's analyzers are implemented using SEC. D-REcommand strings are described as values for *action* identifiers in the SEC configuration file. Table 7-2 shows a list of actions that can be called by SEC.

Table 7-2 Actions Called by SEC.

Actions	Meaning
shellcmd /usr/share/dre-demo/d-script/action/ act_logging.sh	Write the results of analysis to a log
shellcmd usr/share/dre-demo/d-script/action/ act_reboot_sys_container.sh	Restart the system container
shellcmd /usr/share/dre-demo/d-script/action/ act_undo_sys_container.sh	Undo changes to the system container
shellcmd /usr/share/dre-demo/d-script/action/ act_kill_batch_sys.sh	Terminate batch jobs
shellcmd /usr/share/dre-demo/bin/dcase-status	Change the displayed status of D-Case monitor nodes
event	Generate SEC internal events
create, delete	Change the SEC state

REFERENCES

[1] Simple Event Correlator: http://simple-evcorr.sourceforge.net/
[2] D-Case Weaver: http://www.jst.go.jp/crest/crest-os/tech/DCaseWeaver/index-e.html

7.3 ROBOT SYSTEM CASE STUDY

7.3.1 ART-Linux—A Real-Time OS Providing D-RE Functionality

Robot systems for practical applications must be able to execute multiple real-time cycles having specific periods with very little jitter. Accordingly, non-real-time operating systems including Linux are not suited to these applications. Even though preemption latency is gradually improving with successive versions of Linux, the control loop of a typical robot has a period of 1 ms, and jitter must be kept within several percent of this, which is not achievable with Linux. As a solution, various operating systems such as VxWorks and QNX have been designed specifically for real-time processing in embedded systems. Meanwhile, when we want to make a robot system dependable using the DEOS approach by exploiting D-Case monitor nodes at multiple locations within the system, it causes the associated overhead to increase; yet monitoring latency must also be kept as low as possible in order that failure may be accurately detected.

In order to satisfy the constraints associated with real-time processing, we set about developing a real-time operating system called ART-Linux in 1998. Extended from the Linux kernel, this operating system provides real-time system calls for programs in user space, thereby realizing real-time functionality in the order of 10 μs. In order to provide D-RE functionality, we designed and developed ART-Linux so as to be capable of combining, in any ratio, Linux supporting asymmetric multi-processing (AMP) bound to several CPU cores and Linux supporting normal, non-real-time symmetric multi-processing (SMP) using multiple CPU cores. The main advantage afforded by this operating system is that, whereas ordinary open-source software and devices are used on the non-real-time SMP portion, the real-time AMP portion allows dependability functions related to real-time monitoring, real-time control, safety, and dual configuration for reliability to be implemented independently of one another, while communicating through dedicated I/O.

D-System Monitor running on ART-Linux can access the system's shared memory to monitor and record system data provided by sub-systems deployed on separate CPUs, making it an excellent solution in terms of real-time-ness and durability.

Figure 7-19 shows a typical configuration of a system that can be realized using ART-Linux. This example features eight CPU cores (the maximum number currently supported), three of which run normal, non-real-time Linux. The remaining five cores, however, host mutually independent instances of ART-Linux in order to support real-time functionality.

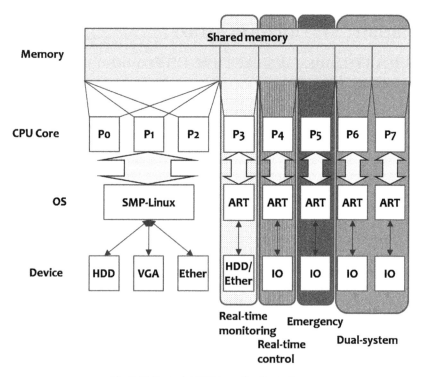

Fig. 7-19 Typical ART-Linux Configuration.

Non-real-time SMP Linux

In the example shown in Fig. 7-19, non-real-time SMP Linux is assigned to CPU cores P0, P1, and P2. Here, we can see that standard devices such as hard disks, monitors, and networks are assigned to this operating system, and its roles are as follows:

- Interfacing with the user.
- Processing for I/O hardware that does not require direct robot control, such as displays and Ethernet equipment; as this processing generates multiple interrupts, it can easily cause problems in real-time operating systems.
- Intelligent robot processes such as long-term planning, map drawing, and recognition based on model matching that do not require real-time execution; the results of these processes are relayed to the real-time control system and others via a virtual network or shared memory.
- Execution of various open source libraries developed without the need for real-time functions.

With the exception of the memory assignment section, this operating system essentially comprises standard SMP Linux, giving it the advantage of supporting software developed for standard Linux without customization.

Real-time monitoring AMP Linux

In the example shown in Fig. 7-19, real-time monitoring AMP Linux is assigned to CPU core P3. This operating system monitors the internal state of the other SMP and AMP systems via shared memory and stores the corresponding logs. When it detects anomalies in the other operating systems in real time, it notifies the emergency operating system, which will be discussed below. The disks used to store logs and separate network devices necessary for the emergency operating system for emergency situations are assigned to this operating system.

Real-time control AMP Linux

In the example shown in Fig. 7-19, real-time control AMP Linux is assigned to CPU core P4, and the I/O devices necessary for controlling real-time devices, such as various types of sensors and actuators, are assigned to this operating system. This system also writes its internal state to shared memory so that the real-time monitoring system can monitor it. Multiple instances of real-time control AMP Linux can be allocated to individual CPU cores as necessary and available. As it can be dedicated entirely to controlling real-time devices with no other overhead, the real-time control AMP Linux operating system provides for excellent real-time control performance with low jitter. This configuration also contributes to easier system design and higher failure tolerance.

Emergency AMP Linux

In the example shown in Fig. 7-19, emergency AMP Linux is assigned to CPU core P5. This operating system is also assigned its own dedicated I/O devices. It detects anomalies in the robot system by receiving notifications from the monitoring operating systems and from other information received via shared memory or network. Upon detecting anomalies, it executes safety countermeasures such as cutting off the supply of power to the motors to effect an emergency stop.

By implementing this emergency operating system in an isolated manner, it is possible to avoid problems such as the other real-time systems blocking the execution of necessary processes. Meanwhile, there are many situations where cutting off the supply of power could actually exacerbate the damage—for example, when a humanoid robot is walking, a vehicle is travelling at high speed, or an arm is carrying a heavy object overhead. Emergency AMP Linux facilitates the calculation functionality required in order to avoid this type of outcome by isolating it from the other operating systems.

Dual-system AMP Linux

Dual operating systems are often used as a means of ensuring reliability, and we now describe how we realize a dual configuration in this case-study system. In the example shown in Fig. 7-19, a dual system is configured by assigning real-time AMP operating systems to CPU cores P6 and P7. Each is assigned its own I/O devices. As systems become ever more complex yet must still function safely in practical settings, dual or even multiple systems will be highly important in ensuring reliability.

7.3.2 Interactive Robot Case Study

In this section, we look at the process of argument-based consensus building among multiple stakeholders using the example of an interactive robot. The robot in question operates on the exhibition floor of Japan's National Museum of Emerging Science and Innovation (the Miraikan), and there, it is required to travel around the facility, verbally interacting with visitors, announcing demonstration times and topics, and assisting in a range of other ways while constantly avoiding people and obstacles. We examine how the robot designers and service providers used a D-Case for five requirements— functionality, operation, safety, accountability, and improvement—to develop an argument and build consensus in all robot-development stages from design to verification.

The floor plan shown in Fig. 7-20 covers an area of roughly 30 x 130 meters of the building's third floor exhibition space. Approximately 10,000 people visit the Miraikan every day.

In terms of hardware, the robot platform comprises a Pioneer 3-DX two-wheel, two-motor mobile base, on which a Velodyne HDL-32E multilayer laser distance sensor, a Microsoft® KINECT sensor, and a microphone array have been mounted. In order to prevent injury, a transparent shield has been placed over the bumper section containing the sensors, the chassis (and its polyurethane foam coating), and the optical sensor section.

Turning to software, Robot Operating System (ROS) middleware is used to integrate the various sensors and software functions, and software modules are implemented individually for recognition, planning, and control. Functionality wise, the recognition module uses the multilayer laser distance sensor to perform position recognition, mapping, and tracking of moving obstacles such as visitors, as well as detecting stationary obstacles and identifying areas in which the robot can safely travel. In addition, it detects stationary obstacles and visitor postures with the KINECT sensor and uses the microphone array for sound-source localization, separation, and recognition. The planning module, meanwhile, is required to plan routes from the robot's recognized map position to the current destination that provide for the quickest travel time while avoiding obstacles and maintaining a high level of safety. In addition, it must also monitor the paths that visitors are taking, and if necessary, plan suitable avoidance routes. The role of the control module is to drive the robot along the planned routes, monitoring relative speeds with respect to stationary and mobile obstacles and adjusting the travel speed as appropriate.

7.3.3 Interactive Robot D-Case

A D-Case was developed in order to assure the dependability of the interactive robot. Dealing with the five requirements of functionality, operation, safety, accountability, and improvement, it comprises a total of 159 nodes—that is, 66 goal nodes, 29 evidence nodes, 28 strategy nodes, 12 context nodes, 7 undeveloped nodes, and 17 monitor nodes (Fig. 7-21).

(1) Argument concerning functionality

The functionality of each of the recognition, planning, and control modules was argued from the perspective of robot services. In the case of the recognition module, the argument covered the following points:

- Accurate mapping of the floor area;
- Accurate position recognition within maps and maintaining position-recognition error within the allowable range even when the robot is

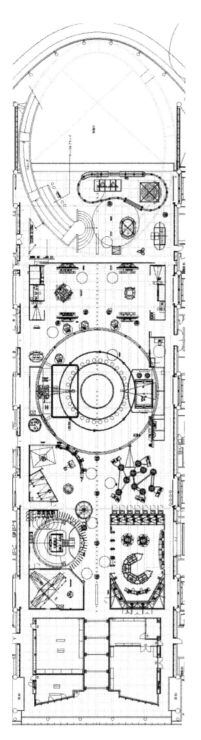

Fig. 7-20 Floor Plan for the Miraikan.

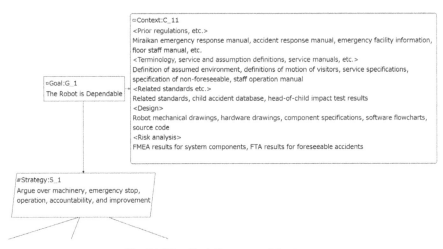

Fig. 7-21 Top Goal, Context, and Strategy.

surrounded by visitors and the sensors' field of view is obscured to a certain degree;

- Defining, accurately detecting, and mapping stationary obstacles, and ensuring that robot behavior will not be adversely affected by the largest possible floor-level difference that cannot be detected;

- Defining and accurately mapping moving obstacles such as visitors, reliably tracking these obstacles even when hidden, and accurately predicting the positions of multiple persons in close proximity to one another; and

- Defining and accurately detecting visitor postures and appropriately processing predictive results for parts of the body not currently visible.

The argument for the planning module focused on the following functions:

- Deriving routes to assigned destinations using acquired map information as input data such that the shortest possible travel time can be realized without compromising safety and the calculation cost can be minimized;

- Setting of cost functions for safety and shortest travel times;

- Setting of new destinations for travel routes when waypoints cannot be reached; and

- Updating map data by adding any newly detected stationary or moving obstacles and online planning of new routes in response.

Last but not least, the argument for the control module covered the following:

- For a given path, determining the precision of robot turning and advancing based on a covariance matrix of the recognized robot position and error with respect to assigned routes, and accurately following the path;

- Keeping a prescribed distance or further from recognized and mapped stationary obstacles; and
- Local avoidance so as to avoid blocking the foreseen paths of recognized moving obstacles.

In the argument concerning these functions, the relevant scientific principles, software specifications, and the like were described as contexts and the results of performance evaluation from actual operating tests were described as evidence, thereby allowing the scope of the argument to be explicitly defined. In order to acquire this evidence, test driving was performed one day a week after the Miraikan was closed at night for one year in the actual operating environment, and all the input and output to and from software was recorded using monitor nodes. In this regard, the multi-core functionality of ART-Linux made it possible to include as many of these nodes as required without compromising system performance. Meanwhile, these monitor nodes are also used during actual operation for accountability and improvement.

As a result of test driving, some improvements were achieved through argument and creating a new version of D-Case—specifically for cases where certain obstacles cannot be reliably observed due to the monitoring frequency being too slow or the obstacles having an optical characteristic close to either total reflection or non-reflection. In such a case, the argument process led to agreement that these obstacles should be designated as non-detectable and that staff should manually add them to the robot's maps in the operation phase. Safety-related issues concerning bumpers and the like were also resolved through bumper redesign. That is to say, we discussed and argued with respect to non-foreseeable failures that became foreseeable with redesign and countermeasures in the operation phase. Figure 7-22 shows the trace of robot movement while it produces a map by recognizing moving and stationary obstacles.

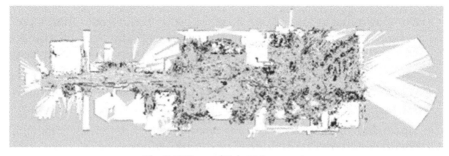

Fig. 7-22 Trace of Robot Movement.

(2) Argument concerning operation

Dedicated science guides and volunteers are assigned to the Miraikan's exhibition floor. The robot system's direct stakeholders comprise these people, the robot operators, and the visitors to the exhibition floor. Decisions concerning the required mode of operation were made based on arguments put forward by these stakeholders, and they included:

- The timing and mode of intervention in order to ensure service continuity when system errors that both can and cannot be foreseen occur in safety-related and function-related areas; and

- Response to failure.

In terms of scope, the operation manuals were produced according to the context descriptions, and they were referenced in order to ensure consistency with the D-Case. In addition, specifications for preemptive system response with respect to this type of intervention by staff were argued, and it was agreed that the results of actual operation tests would be added as evidence.

Of particular importance in terms of operation is a procedure for improvement in the *Change Accommodation Cycle* when an unforeseeable incident has occurred or continuity of service is not possible. This issue was also addressed in the argument process and reflected in the manuals.

(3) Argument concerning safety

The robot system's safety-related argument concerned passive and active safety, and in terms of passive safety equipment, there was particular focus on the system stopping when collision could not be avoided. This is made possible by a collision detection system using bumper switches, motor current sensors, and encoders, which are implemented as hardware isolated from the robot's information system. Assuming that power is being supplied, the robot can safely stop without causing damage if it should make contact with an object. In addition, elastic buffer material is also used to minimize any injury if it should collide with a person.

Context for this argument took the form of the functional-safety standard, robot-safety regulations, foreseeable and unforeseeable risk (such as somebody falling onto the robot), studies concerning collision with the head of a child and associated injury, and so forth. The results of FMEA, test results to determine whether the robot system can operate as designed, and buffer-material hardness measurements were added as evidence.

(4) Argument concerning accountability

In order to demonstrate that a system is operating normally, or for example, that some kind of unforeseeable anomaly has occurred, input and output data for each of its elements must be recorded in real time and stored. Using D-RE functionality made possible by ART-Linux, simply copying monitor node data to shared memory allows a logging system running on a different CPU core to store that data. In this way, data can be recorded whenever necessary.

In this robot system, many different events are monitored and tens of megabytes of compressed data are logged every second. In the event that an anomaly is detected in any element of the system or within the related external environment, a data collection system configured for the corresponding monitor node will issue an alert and the system will safely transition to a stopped state. This corresponds to the DEOS *Failure Response Cycle*. In addition, records of each of these actions are used to identify the nature and location of the anomaly, thereby facilitating transition to the *Change Accommodation Cycle*.

In order to make this possible, the argument process focused on what types of anomaly could be foreseen and logged. As a result, it was agreed that a total of 17 monitor nodes were to be added for the input and output data of all sensors and motors, the input and output data of all functions, and the control input from the operator's console. In addition, test results showing that anomalies could be properly logged were added as evidence.

(5) Argument concerning improvement

It was agreed that services should be halted and the *Change Accommodation Cycle* initiated whenever service continuity could not be achieved on-site by the operator alone in the event of, for example, an unforeseeable accident involving the robot system, anomalous behavior not resulting in an accident, or malfunction of a sensor, motor, or other sub-system. After individual arguments were made concerning malfunction, foreseeable failure not planned for in design, and unforeseeable failure, records from the above-mentioned monitor nodes would be used to reproduce the failure and the system would be improved. In regard to unforeseeable failure, meanwhile, it is unclear whether the monitor nodes have been correctly arranged. An attempt was made to reproduce such a failure and implement improvements based on sensor records, transcripts of interviews with staff involved in the follow-up, and the like.

7.3.4 Summary

In this section, we looked at an interactive robot operating on the exhibition floor of the National Museum of Emerging Science (the Miraikan) as a D-RE case study. In specific terms, we discussed arguments concerning five requirements governing the robot's *Failure Response Cycle* and *Change Accommodation Cycle*—namely, functionality, operation, safety, accountability, and improvement. And in terms of a robot configuration that satisfies these requirements in practice, we looked at the D-Case developed in order that designers and operators could reach consensus as stakeholders. Describing the related argumentation in detail would involve discussion of specific components such as sensors and actuators, as well as standards and regulations. To do so would be beyond the scope of this case study, so we have presented just an outline. Nevertheless, this should suffice to show that in-depth arguments involving scope, regulations, evidence, design documents, test results, manuals, and so forth are made for such a system.

The D-Case arguments will be constantly updated over the full lifecycle of the system, from design right through to actual operation. With the monitor nodes recording operation records on an ongoing basis, arguments concerning failure or updates can be made at any time.

We confirmed that consensus building on the basis of D-Case arguments offers the following advantages:

- Clarification of the overall scope;
- Identification of conditions for initiating the *Failure Response Cycle* and *Change Accommodation Cycle*;
- Clarification of material required to support the argument;
- Integration of risk analysis results from FTA, FMEA, and the like;
- Integration of compliance evaluation in terms of existing safety and technical standards;
- Identification of issues associated with each goal and recording of arguments concerning agreement conditions;
- Clarification of accountability information acquired using monitor nodes;
- Identification of manuals for operation, maintenance, and so forth;
- Identification of foreseeable failure not planned for in design;
- Unambiguous definition of terminology and parameters; and
- Clarification of conditions relating to precision and the like.

It should be noted, however, that D-Case developers must still be responsible for determining whether the argument is complete, whether overall consistency is maintained following partial revisions, whether the stakeholders' agreement with each of the goals is consistent, whether subordinate arguments are compatible with the top goal, and so forth. Some of these issues are resolved by D-ADD, which is described in Chapter 9, but appropriate care is needed.

In the robot system of this case study, transition from the *Ordinary Operation state* to the *Change Accommodation Cycle* is initiated by staff members. All actions within the *Failure Response Cycle* have been implemented using ordinary (application) programs, although they should have been described in D-Scripts for more flexible control. Figure 7-23 shows the robot in operation at the Miraikan.

Fig. 7-23 Robot in Operation at the Miraikan.

REFERENCES

[1] Tokoro, M. (ed.). 2012. Open Systems Dependability—Dependability Engineering for Ever-Changing Systems, CRC Press.
[2] Matsuno, Y., T. Takai and S. Yamamoto. 2012. An Introduction to D-Case: Let's Write Dependability Cases, ISBN 978-4-86293-079-8 (in Japanese).
[3] Matsuno, Y., J. Nakazawa, M. Takeyama, M. Sugaya and Y. Ishikawa. 2010. Towards a Language for Communication among Stakeholders. In Proc. of the 16th IEEE Pacific Rim International Symposium on Dependable Computing (PRDC'10), pp. 93–100.
[4] D-Case Editor: http://www.dcase.jp/editor_en.html
[5] DEOS Project White Paper Version 3.0, DEOS-FY2011-WP-03J.
[6] ART-Linux: http://sourceforge.net/projects/art-linux/
[7] Kagami, S., Y. Ishiwata, K. Nishiwaki, S. Kajita, F. Kanehiro, W.N. Yoon, N. Ando, Y. Sasaki, S. Thompson and T. Matsui. 2012. Design and Implementation of ART-Linux Capable of Using Multiple CPU Cores by Combining AMP & SMP in Proceedings of 17th Robotics Symposia, pp. 521–526, Hagihonjin, Hagi City, Yamaguchi Prefecture, March 2012 (in Japanese).
[8] Ishiwata, Y., S. Kagami, K. Nishiwaki and T. Matsui. 2008. Design & Development of ART-Linux 2.6 for a Single CPU, Journal of the Robotics Society of Japan, Volume 26, Edition 6, pp. 546–552, September 2008 (in Japanese).

7.4 SECURITY

7.4.1 Security in the DEOS Process

The open systems we envisage for the DEOS Process are accessed by an undeterminably large number of users and comprise software components from different vendors running in an interdependent fashion. When compared with closed systems, therefore, it is much more difficult to foresee the types of security threat to which they will be exposed. Accordingly, it is of vital importance that security mechanisms be put in place in these open systems in order that security incidents caused by unforeseeable and unknown threats can be dealt with as quickly as possible. In other words, a framework capable of assuring security even with respect to unforeseeable threats is needed.

That said, unforeseeable threats cannot easily be eliminated in advance, if only because it is practically impossible to know what countermeasures will be required. In other words, we cannot apply the same technique as often used for foreseeable threats, where all known risks are identified and the countermeasures implemented for each of those individually are subjected to a process of verification.

The DEOS approach does not apply that technique—instead, it is assumed that security may have been compromised whenever system operation deviates from expected norms. Rather than focusing on what could have caused the breach, the fact that performance of a system is not at the required level is seen as sufficient evidence of a potential security issue. Conversely, if the system is operating as designed, it is assumed that security has not been compromised. Consider, for example, a system designed and implemented such that a user *A* cannot access information *X*. If a situation arose where user *A* could access that information, some kind of security breach would be assumed to have occurred. The cause of the breach is not the focus of attention.

In line with this approach, the following three mechanisms for assuring security were proposed during the course of the DEOS Project:

(1) A mechanism for verifying whether systems employing the DEOS Process are operating as designed: Returning to the above example, attempts could be made at regular intervals to access information *X* using user *A*'s authorization, thereby determining whether or not the system is performing normally. This mechanism has actually been employed to assure operating system security within the DEOS Process—specifically, it is integrated into D-System Monitor, and more details are provided in Subsection 7.4.3.

(2) A mechanism for restoring a system that is not operating as designed due to some kind of breach to a sound operating state: The methods for restoring normal operation must be determined based on the specifics of the security breach, the nature of the application where the breach occurred, and other similar factors. And for this reason, D-RE alone

cannot select the appropriate recovery methods. Instead, it must provide the various primitives required for recovery, and the most appropriate primitive must then be selected by D-Scripts. Specifically, D-Scripts would select recovery primitives on the basis of monitoring information acquired from D-System Monitors. A wide range of recovery functions could conceivably be integrated into D-RE, but at the present point in time, reboot primitives are provided on account of the relatively broad applicability of this approach. More details are provided in Subsection 7.4.4.

(3) Automatic update functionality for repairing the actual vulnerabilities: As described in Subsection 7.4.2, various types of vulnerabilities are caused by faults or bugs in the software. Whenever any type of vulnerability is detected, therefore, a patch must be applied to fix the fault or bug that caused it, and an automatic update function makes this possible without having to take the information system offline or stop any of the software running on it. More details concerning this mechanism are provided in Subsection 7.4.5.

The DEOS Project has attached the highest importance to the assurance of operating system security. Operating systems provide a platform on which software can run, and if vulnerability in this platform were to be targeted and illegal code introduced, the operating system itself could no longer be trusted. Regardless of how well the system has been equipped with access restriction mechanisms, data encryption functionality, and the like, they would all be rendered useless in such a situation. In order that the security of an entire information system may be assured, therefore, it is critical that its operating systems can be guaranteed to be operating as originally designed and implemented. From this perspective, it is important to focus on vulnerabilities in the operating systems—that is, faults that could become a security hole and lead to system failure—instead of applying access control models and other similar techniques. Without a framework for guaranteeing that operating systems can run as designed, we can no longer trust the access control, data encryption, and other security functionality that they provide, and the entire information system will be in danger of collapsing.

Much more so than with closed systems, it is highly important that open systems can deal with threats of an unknown nature. One of the key features of DEOS Project research was to provide a highly versatile detection mechanism with little dependence on specific attack patterns or methods by identifying *deviation from expected behavior* as indication of a possible security anomaly. Thanks to this approach, it is possible to detect many different classes of malware, such as key loggers, rootkits, adware, and fake antivirus software. In fact, no other comparable approach can detect such a wide range of malware.

In order to apply this proposed approach, the behavior expected of an operating system in specific situations must be defined. However, this may not be possible for all conceivable situations, meaning that susceptibility to illegal code would be undetectable in certain cases. As a result, the approach cannot be applied in situations where operating system behavior caused by malware is indistinguishable from normal behavior. Any malware that could cleverly conceal its operation behind normal operating-system processes would thus pose a threat to our approach. We must keep a watchful eye out for any technical developments that would make it possible to create this type of malware.

7.4.2 Vulnerabilities & Faults

Another distinctive feature of the DEOS security framework is, in a broad sense, considering vulnerabilities to be faults and security incidents to be one specific manifestation of failure. It may be said that programming errors—or in other words, faults—are the cause of serious attacks such as operating-system hijacking. Faults that constitute security holes are extremely difficult both to detect and to remove in advance through testing and verification. There have, for example, been many cases in the past where integer overflow has led to the development of serious security holes. We often see code of the form $p = kmalloc(n * size);$ used for memory allocation in the kernel. If the value of n is too large, the result of the $n * size$ calculation will cause an overflow, and the amount of assigned memory will drop, leading to a buffer overflow. Symptomatic types of verification method can be developed for this type of fault, but given that new incidents are being reported all the time, this approach would not seem to be adequate.

Unfortunately, Linux and most other modern operating systems include a great many faults. Operating systems subjected to rigorous verification such as the seL4 microkernel are now available; however, this is an embedded operating system with limited functionality resulting in a smaller code size that makes verification easier. On the other hand, Linux, which is used in a vast array of different applications from embedded systems to super computers, is huge in size and extremely complex. This makes formal verification all but impossible in practice. As a result, this type of verification is either totally abandoned or used for specific sub-systems only.

Figure 7-24 shows how the number of Linux faults has trended over time [6]. It should be noted that these graphs only count NULL pointer references, memory allocation size errors, and other problems that are relatively easy to discover by static analysis: they do not include race hazards, memory leaks, and other faults that are caused by a complex combination of conditions. Yet despite this limited scope, the top graph clearly shows that Linux—regardless of version—has continually averaged around 700 bugs.

In order to assure a high level of operating-system security, it is crucial that breaches can be detected and normal operating conditions can be rapidly restored, even in the case of hijack attempts that exploit vulnerabilities. It is not easy to identify those kernel faults that could result in a security hole.

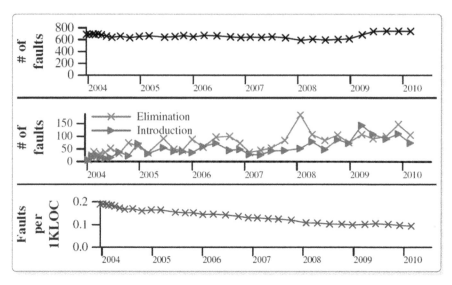

Fig. 7-24 Number of Faults in Linux Over Time.

So rather than debating whether each individual fault in the kernel could constitute such vulnerability, we believe that countermeasures should be implemented with the understanding that all kernel faults could result in a security hole, and furthermore, we feel that this represents a highly practical solution. This viewpoint also equates faults with vulnerabilities and requires them to be addressed via the same mechanisms. D-RE deals with security-incident detection and countermeasures on exactly the same basis as normal failure detection and countermeasures. In the next three sections, we look at the DEOS security mechanisms built into D-RE.

7.4.3 Mechanism for Illegal-Code Detection

D-System Monitors and D-Visors are utilized in the detection of illegal code embedded in operating system kernels. To this end, a D-Visor conceals the hardware and implements the D-System Monitor runtime environment. The D-System Monitor is isolated from the operating system being monitored, and its calculation resources cannot easily be accessed from the monitored VM. The operating system in question can be monitored from this environment (which is actually a different virtual-machine environment from that of the operating system) in order to investigate whether or not it is operationally sound (Fig. 7-25). The data produced by this investigation must have been processed using D-Scripts, and based on the results thereof, the appropriate recovery measures can be taken—in specific terms, this is achieved by initiating Phase-based Reboot and ShadowReboot as described below.

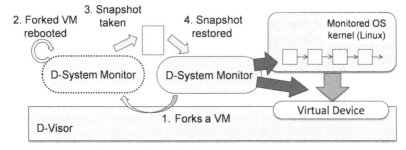

Fig. 7-25 D-System Monitor and D-Visor.

D-System Monitors keep a careful eye on operating system behavior in order that the *Failure Response Cycle* can run smoothly. In combination with a D-Visor, each one can accurately monitor privileged-register access, the execution of privileged instructions, input, and output by the operating system in question, as well as the content of these processes. In addition, a D-System Monitor can also generate interrupts in the operating system being monitored or software interrupts that launch system calls in order to observe how the operating system behaves in response. Using this approach, we can guarantee that an operating system is operating as intended, or in other words, is operationally sound.

For example, FoxyKBD—one function implemented in the D-System Monitor—provides a mechanism for detecting key loggers maliciously embedded in an operating system. FoxyKBD can simulate high-speed keyboard input at levels that would be impossible for humans so as to amplify the actions of any key logger. For example, a key logger receiving a large volume of keyboard input must execute a correspondingly large amount of disk I/O in order to record that data. If a statistically significant correlation is seen between the simulated keyboard input and the resulting increase in disk I/O, it can be determined with a high degree of certainty that a key logger is present. The majority of existing key loggers can also collect keyboard-input data at regular intervals, and therefore, timer interrupts could be simulated to accelerate artificially the passage of time in the monitored operating system, thereby allowing this type of malware to be detected. This technique has actually been applied to 56 different key loggers collected via the Internet, and regardless of type, they could all be detected.

The chief advantage of this technique is that fixes do not need to be implemented for each different malware type. Instead, malware can be classified by the type of malicious action it takes, and fixes can be implemented on a class-specific basis. For example, malware that steals keystrokes and malware that tampers with file metadata can be assigned to different classes, and monitoring methods can then be provided for each class. In this way, general-purpose detection mechanisms not tied to any specific malware type, attack vectors, or the like can be configured, thereby ensuring a high level of resistance, even to unknown threats. For example, to detect malware that conceals the existence of files, a list of file names acquired from the result of I/O processing could be compared with a list created from system-call return values in order to identify any inconsistencies.

7.4.4 Mechanism for Illegal-Code Recovery

The best way to prevent security incidents is to eliminate vulnerabilities from all software layers before they can be exploited. That said, as OS kernels are required to deliver higher and higher levels of performance and grow ever more complex, it is practically impossible to fully remove all bugs that cause vulnerabilities during the development process. Accordingly, we need a method for mitigating the effect of these bugs in the system's operation phase. The necessary routines can be called based on analysis of the state of the system by D-Scripts.

It is well known that the adverse effect of bugs in an operating system kernel can be mitigated by rebooting. In the event of a kernel crash or memory leak due to a bug, rebooting the kernel will restore normal operation without having to identify the specific cause of that bug. Meanwhile, many kinds of malware that target vulnerabilities in an operating-system kernel reside in memory in order to remain concealed. Rebooting the kernel also resets the content of memory, and thus is an effective way to wash out this type of malware. It should be mentioned, however, that whenever an operating system is rebooted, so must the applications running on it, and this can lead to extended downtime.

D-Visors provide support for Phase-based Reboot [1]—a high-speed reboot-based recovery method that targets reboot phases. In contrast to operating-system reboot when, for example, software has been updated, rebooting in order to recover from a failure executes the same phases as normal system booting. Accordingly, the state immediately after the most recent boot can be restored to produce the same effect as a full operating-system reboot.

As shown in Fig. 7-26, the state of an operating system immediately after it is booted can be stored and then restored when required in order to return to normal operating conditions, achieving the same result as a full operating-system reboot, but in a much shorter time. Snapshot functionality is used to store the state of the system; however, simply restoring a snapshot would also set the hard disk back to its prior condition. As a solution, the current disk is read immediately after the snapshot is restored, the *FileSystemObject* in memory is suitably updated, and the updated disk condition is propagated to the system. This approach has been shown to actually reduce downtime associated with operating system rebooting by up to 93.6%.

In addition, certain modifications have been made so that soft-state memory areas, such as the page cache, are not stored as a snapshot in order to ensure that operational soundness can be rapidly restored even when the

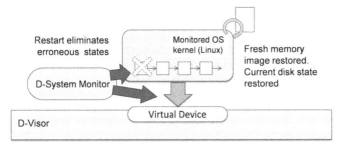

Fig. 7-26 Phase-based Reboot.

amount of memory consumed by the virtual machine is large. The D-Visor also incorporates mechanisms for speeding up its own reboot-based recovery [2, 3] and for preventing any drop in performance upon reboot [4].

7.4.5 Action after Dynamic Update

Whenever fixes have been made to an information system after the occurrence of a failure, the system in question normally needs to be rebooted. A typical example of this would be rebooting after applying a patch to the operating system kernel or an application, and the addition of new security functionality to a D-System Monitor is no different. Adding a new failure countermeasure thus means that services and monitoring will have to be temporarily halted. But as bugs grow in number in line with ever-increasing software complexity and modes of malicious attack also become more ingenious, fixes must be implemented on a frequent basis.

The D-System Monitor and the D-Visor have been designed with this in mind, and their architecture facilitates easy and efficient implementation of fixes. Many techniques for dynamic updating of operating system kernels and applications have been developed, and these too can be incorporated. In the past, rebooting has been unavoidable in cases where dynamic updating techniques cannot be applied; however, the D-Visor's high-speed updating mechanism provides for smooth implementation of failure countermeasures.

D-Visors also provide a method for reducing downtime associated with software updates known as ShadowReboot [5]. This is achieved by concealing operating-system reboot actions from active applications. More specifically, a virtual machine with the same state as the operating condition is spawned whenever a reboot is required, and the operating system reboot is executed on this virtual machine. Meanwhile, the applications continue to run on the original virtual machine. Once the reboot sequence has ended, a snapshot of the spawned virtual machine is taken, and this snapshot is restored on the original virtual machine in order to effect the actual software update (Fig. 7-27). The updated content of virtual disks is also retained when restoring the snapshot, meaning that a normal update can be completed with considerably reduced

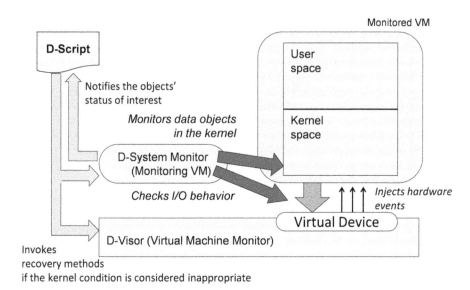

Fig. 7-27 ShadowReboot.

downtime. Furthermore, this approach applies not only to monitored virtual machines that are providing services, but the D-System Monitor performing the monitoring can also be updated. Prototype implementation carried out to date has shown reductions of between 83% and 98% in downtime associated with updates.

7.4.6 Summary

As applications and hardware evolve technologically, operating system kernels are ballooning in size and becoming ever more complex. The upshot is that kernels continue to include many programming bugs and are increasingly the targets of malicious attack, despite being required to deliver high levels of reliability. These factors can cause services to be interrupted, and in the worst case scenario, the services themselves may be tampered with or data stolen. In this section, we looked at a runtime framework premised on operating-system kernel failure. One of the key features of the approach in question is to provide a highly versatile detection mechanism with little dependence on specific attack patterns or methods by identifying *deviation from expected behavior* as indication of a possible security anomaly. This allows for the detection of many different classes of malware, such as key loggers, rootkits, and the like. In order to apply this proposed approach, the behavior expected of operating systems in specific situations must be defined; however, this is not possible for all conceivable situations, meaning that susceptibility to illegal code would be undetectable in certain cases. As a result, the approach cannot be applied in situations where operating system behavior caused by malware is indistinguishable from normal behavior. Any malware that could cleverly conceal its operation behind normal operating system processes would thus pose a threat to our approach, and therefore, we must watch vigilantly for any technical developments that would make it possible to create this type of malware, and if necessary, develop the required countermeasures.

REFERENCES

[1] Yamakita, Y., H. Yamada and K. Kono. 2011. Phase-based Reboot: Reusing Operating System Execution Phases for Cheap Reboot-based Recovery. In Proc. of the International Conference on Dependable Systems and Networks, pp. 169–180.

[2] Kourai, K. and S. Chiba. 2007. A Fast Rejuvenation Technique for Server Consolidation with Virtual Machines. In Proc. of the International Conference on Dependable Systems and Networks, pp. 245–255.

[3] Kourai, K. and S. Chiba. 2011. Fast Software Rejuvenation of Virtual Machine Monitors, IEEE Transaction on Dependable and Secure Computing, Vol. 8, No. 6, pp. 839–851.

[4] Kourai, K. 2011. Fast and Correct Performance Recovery of Operating Systems Using a Virtual Machine Monitor. In Proceedings of the ACM International Conf. on Virtual Execution Environments, pp. 99–110.

[5] Yamada, H. and K. Kono. 2011. Traveling Forward in Time to Newer Operating Systems using ShadowReboot. In Proc. of the ACM Asia-Pacific Workshop on Systems, pp. 12:1–12:5.

[6] Palix, N., G. Thomas, S. Saha, C. Calves, J. Lawall and G. Muller. 2014. Faults in Linux: Ten year's later. In Proc. of the International Conference on Architectural Support for Programming Languages and Operating Systems, pp. 305–328.

8

D-Script—Support for System Operation based on D-Case Agreements

The DEOS Process makes it possible to enhance the dependability of information systems in line with stakeholders' agreements throughout their lifecycles and at all individual stages thereof, such as design, development, modification, and operation. Against this backdrop, D-Script provides a means for executing operational procedures that the stakeholders have agreed upon in D-RE.

Within the DEOS Process, D-Case is used to make arguments concerning dependability, and the stakeholders' agreements reached as a result are managed as digital data. Each D-Script is described in the D-Case notation as an *action* that the system in operation must carry out, and together, they form one part of the agreement. These actions are sent in the form of executable scripts to D-RE, where they are executed by D-Script Engine. The success or failure of execution is then fed back to the process as evidence.

8.1 PROGRAMS AND SCRIPTS

The DEOS Architecture constitutes a runtime environment model abstracted for systems providing services on the basis of the DEOS Process. Applications developed in line with agreed functional requirements are used to provide the services in question. However, the processes required for delivering these services are normally executed in an open environment comprising multiple, interlinked applications, and therefore, it is practically impossible to fully satisfy all dependability requirements, regardless of how well the individual applications have been developed.

D-Scripts provide a means of implementing additional dependability requirements during system operation. In contrast to application programs, they are not provided in the form of services defined based on functional requirements; instead, D-Scripts serve as a support tool for operators, managing the lifecycles of applications, protecting data, balancing the loads of multiple applications, and so forth. In this capacity, they allow the applications to execute in such a way that dependability requirements can be satisfied.

8.1.1 Significance of Stakeholders' Agreement on Scripts

Scripts have, since before the invention of time sharing systems, been used in the form of batch files in order to support operation. These days, a wide variety of scripts are used in all types of computer systems from embedded to enterprise. In Linux, the mainstream operating system for building computer services, typical examples are the Bourne shell, extended shells (such as csh and zsh), and also Perl and Python for greater programming capabilities. Meanwhile, major cloud computer platforms such as Amazon Web Services also utilize scripts to control system operation.

Operation scripts were first put to use in the automation of operation-related procedures that had previously been performed manually. This type of script is not developed in the same way as conventional software: the analysis of requirements and signs of failure is often skipped, and testing may well be inadequate. As a result, script malfunction can often lead to system failure and may aggravate the damage done. For example (see Appendix A.3):

- A script used in a stock exchange system for booting up a backup server did not correctly provide notification of the completion of booting. The system assumed that execution of the script had ended in failure, so it booted up a different backup server, causing services to be interrupted.

- An operator of a cloud storage system accidentally launched a personal-information deletion script that he had created but was not approved. The script deleted more than 90% of the files stored by the service, including backup versions.

Given that operation scripts are put to use in the parts of an operational system directly involved with satisfying dependability requirements, their failure to operate correctly, even if only temporarily, can exacerbate the damage caused by system failure. Despite the considerable effect they can have on a system, today's operation scripts are not subjected to quality control in the software engineering sense of the word, and they are often written and executed at the discretion of individual operators.

We developed D-Script to ensure that this type of operation script can be managed on the basis of stakeholders' agreements, that suitable failure and lifecycle management can be applied, and that as a result thereof, system operation based on those agreements can be supported.

Figure 8-1 shows a conceptual view of the overall D-Script system. Here, we can see that the system is split into the consensus building side and the operation side. On the consensus building side, stakeholders reach consensus concerning operation procedures based on the D-Case [1, 2], and planned actions that must be taken in order to satisfy dependability requirements are added to the script. These agreement descriptions are securely stored in D-ADD. Meanwhile, applications run on D-RE, which includes D-Script Engine, on the operation side. The agreed actions are executed by D-Script Engine in order to control these applications, and it stores results of execution in a D-Box as log records. These records are then returned to the agreement side, where they serve as evidence for further development of the D-Case argument. They are also used to improve operation procedures and as the basis for change accommodation.

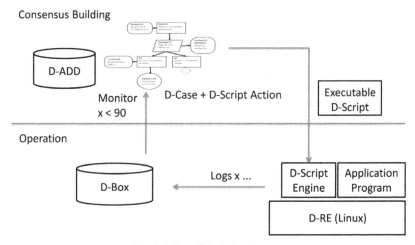

Fig. 8-1 Overall D-Script System.

8.1.2 Role of D-Script in the DEOS Process

Within the DEOS Process, operators have an important role to play in achieving dependability. D-Scripts are described as machine-executable operational procedures in order to automate the tasks that they normally perform manually. Human error on the part of the operator is an extremely common cause of failure, and generally speaking, dependability can be improved by replacing manual operational procedures with machine-executable scripts and confirming in advance that they perform as intended.

In terms of the DEOS Process [3], D-Scripts are used for procedures aimed at satisfying dependability requirements (i.e., non-functional requirements). Consider, for example, a D-Case containing the goal "Data consistency can be assured". In order to achieve this goal, operational procedures for backing up data must be carried out, and D-Scripts would be written for any such backup procedures that could be executed automatically. In this way, D-Scripts always originate from dependability requirements.

In specific terms, the main roles of D-Script within the DEOS Process are as follows:

(1) It gathers system operation data in place of an operator. Data gathering is performed through D-System Monitor and D-Application Monitor as designated by D-Script, not only for constant operation monitoring, but also for system verification when performing in-depth diagnosis in the event that regular monitoring detects an anomaly, and so forth.

(2) It configures the system, increases or decreases available computer resources, reboots the system, updates system tools, cancels updates, and executes other similar procedures normally performed by an operator.

Operation scripts make it possible to combine the above roles, ensure that the tasks in (1) and (2) are executed without fail, and provide for further interlinking. Meanwhile, these scripts also support the execution of sequences of highly-complex procedures and operations that could not be performed in the required integrated manner at human response speeds. Figure 8-2 shows a conceptual view of D-Script operation.

It should be noted that the DEOS Process does not aim to replace all operator activity with D-Scripts. For example, human operators will still be required to make important decisions and implement countermeasures manually in the event of unforeseen circumstances. That said, conditions for calling operators must be agreed upon in advance and rendered in the form of D-Scripts.

Within the *Failure Response Cycle*, D-Scripts are not used to modify application software, add new functionality, or implement new modes of operation. When, for example, services must be degraded in response to a failure, this is achieved using service degradation functionality that needs to have been implemented in the system's applications in advance. During the design of these applications, therefore, operational procedures and other related requirements must be fully analyzed, functionality for supporting the required operation must be implemented, and the corresponding APIs must be made available to D-Scripts.

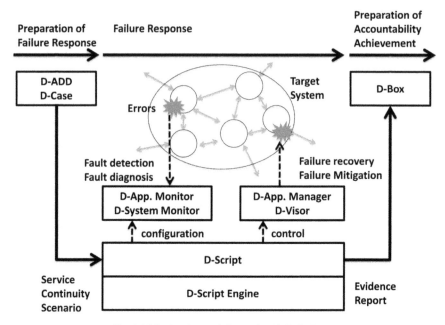

Fig. 8-2 Monitoring and Control with D-Scripts.

Similarly, when application-specific internal information must be acquired using D-Scripts, monitor points for doing so need to have been configured in advance. D-Scripts cannot be used to overwrite applications and setup new monitor points. Once monitor points have been implemented, however, they can be turned on or off as needed depending on when monitoring data is required.

If a specific application did not have adequate functionality for service degradation or the necessary monitor points for referencing by D-Scripts, modifications would be made during a *Change Accommodation Cycle*.

8.1.3 D-Script Engine

Running on D-RE—the DEOS runtime environment for applications—D-Script Engine is a script processor used to execute D-Scripts. As a first step in this process, D-Script Engine gets executable scripts from D-ADD whenever stakeholders' agreements are updated. It then executes these scripts as required, and by storing the results of execution in a D-Box as log records, ensures that evidence can be fed back to the D-Case. We have defined the following three synchronization methods for D-ADD and D-Script Engine:

- Pull synchronization: D-Script Engine regularly checks for updated information in D-ADD. This approach is effective when synchronizing with an indeterminate number of computer systems, such as embedded systems. There will, however, be a time lag before the updated stakeholders' agreement (i.e., the updated D-Case) and its content can be reflected in the operational system.

- Push synchronization: D-ADD sends D-Case contexts and action-function definitions to D-Script Engine and forces synchronization. Time lag is short with this approach, and the behavior of individual D-Script Engines can also be synchronized easily. That said, the target D-Script Engines need to be accurately determined and a means of communication provided. In addition, high-performance security countermeasures are needed to prevent D-Case tampering and the like.

- Integrated synchronization: When D-ADD storage and applications are integrated, they may be placed in the same location.

Within D-RE, D-Script Engine executes as a program with the same authorization as the software it is controlling, and it needs no special permissions. D-Scripts control applications using the APIs made available by D-RE; however, they do not make direct changes to the corresponding executable images.

The processes that D-Scripts can execute are limited by the APIs made available by the D-RE on which the corresponding D-Script Engine is running. For example, the Linux D-RE of the reference DEOS implementation provides for monitoring and lifecycle control on an individual application basis. If, however, a D-RE were to be implemented using the Amazon Web Services cloud platform, it would only be possible to perform these tasks on an operating-system image basis.

Meanwhile, certain APIs are also required for D-Script execution, and those made available by D-Script Engine are as follows:

- API for storing the execution results of D-Script actions in a D-Box

- API for operator notification when machine-based D-Script processing is not possible

These APIs could also be made available in the form of a D-RE library. As long as the above conditions are satisfied, bash, perl, and other common scripting languages could be applied as a D-Script Engine execution platform. Meanwhile, the DEOS Association[1] (Appendix A.2) has made a D-Shell-based,

---

[1]DEOS Association is an abbreviation of Association of Dependability Engineering for Open Systems.

highly reliable D-Script Engine implementation available as an open source solution within the reference D-RE implementation. More details on this engine are provided in Section 8.4.

8.1.4 D-Script Security

The execution of operation scripts can interrupt services and damage data, and as such, it typically poses a security risk. D-Script Engine extracts scripts from a D-Case stored in D-ADD, which is a script source that can be trusted. Nevertheless, scripts stored as editable text files within the operational system could be overwritten by operators. Therefore, there still remains a risk of operation not in accordance with the stakeholders' agreements (even though all modifications are automatically recorded).

In consideration of the security risk associated with D-Scripts, it is necessary to operate D-ADD in a fully secure manner and also to protect the D-Cases stored therein from tampering. Furthermore, only authorized operators must have access to the operational system. If any of these preconditions are not met, it will no longer be possible to assure security at the D-Script level.

D-Script provides an optional script obfuscation function using static single assignment (SSA) form conversion so that the same low level of readability as binary code can be achieved without affecting operation in any way. This means that an operator attempting to change a script would essentially need to do so at the binary code level, which given the difficulty thereof, provides added protection for operation in accordance with the stakeholders' agreements. In the event of unforeseen failure, operators authorized in advance may need to implement emergency countermeasures on-site. In this type of situation, operation can be adjusted in line with dependability requirements and action types.

8.2 D-SCRIPT LANGUAGE DESIGN

Scripts have long been put to use in system operation. In practice, even if a script's source code were to be readable in text format, it would normally take specialist programming knowledge to know, for example, what dependability requirement goal it is associated with. D-Script employs an abstracted notation structure that allows all stakeholders—regardless of their level of technical expertise—to reach agreement on system operation, and in particular, procedures for assuring dependability.

8.2.1 The D-Script Pattern

Our analysis of operation scripts used for monitoring, backup, maintenance management, failure recovery, and so forth in a wide range of applications—

from security camera systems to cloud services—has shown that the vast majority are expressed in the form of (1) an anomalous or undesirable system condition, and (2) the action to be taken in response. None are presented simply as an action. Table 8-1 shows the pairs of undesirable situations and actions in the operation scripts for the Aspen e-learning system, which we developed and has been in use since 2012.

Table 8-1 Undesirable Condition–Action Pairs for Aspen D-Scripts.

	Undesirable condition	Action
1	Service stopped	Boot program
2	Memory leaks	Reboot program
3	Traffic increase	Increase web gateways
4	Data deletion	Backup
5	Action failure	Notify administrator
6	Other error	Notify administrator

In terms of D-Script, an anomalous or undesirable system condition that could precede actual failure is referred to as a *sign of failure* (SOF). The D-Script pattern thus takes the form of an SOF and the corresponding action required to restore normal operation.

D-Script pattern: *SOF*⇨ *Action for restoring normal operation*

In this regard, an SOF is present when the operational system shows actual signs of a foreseen anomalous condition or the potential occurrence thereof.

8.2.2 Symbolizing SOFs

Computer systems contain so many different parameters that the number of potential system conditions is essentially infinite. If an SOF was expressed as, for example, "garbage collection frequency of 45 times per second or greater and simultaneous operating-system swapping at 200 pages per second or greater", it would be difficult to develop the argument and determine what type of action to take in response. Accordingly, D-Script uses language symbols to identify SOFs. In much the same way as doctors use disease names to describe medical conditions, D-Script identifies SOFs using symbolized names such as "Memory leaks".

One might question the actual meaning of "Memory leaks" or what specific condition it implies, but strict definitions can be put in place using D-Case vocabulary functionality in line with the Semantics of Business Vocabulary and Business Rules (SBVR) or other technical standards. Meanwhile, because these symbolized names are ultimately converted into machine-executable scripts, suitable meanings can be assigned at this time—for example, "Memory leaks" could be defined in terms of the results of a diagnostic program.

8.2.3 Faults, Errors & SOFs

Dependability engineering theory classifies system failure mechanisms in line with three different levels—fault, error, or failure [4]. Faults are events that could lead to an error, while errors are defined as anomalous conditions that the system reaches after deviating due to a fault. Failure is defined as the state of the system in which it cannot continue to provide services. As shown in Fig. 8-3, failure occurs in a system when an error is left unaddressed, ultimately manifesting as suspension of services. In addition, any failure can give rise to a new fault, which in turn can produce new failures.

Fig. 8-3 Model of Fault, Error, and Failure.

In terms of countermeasures, attention must be focused not on the system failure itself but the faults that caused it. In this way, measures such as fault forecasting, fault removal, fault tolerance, and fault prevention can address all faults.

Applying a strict dependability engineering interpretation, the D-Script pattern should really define action to be taken in response to a fault [5]. However, "SOF" has a broader meaning than "fault", and for the following reasons, we elected to use SOFs in the D-Script pattern instead of faults:

- D-Case and D-Script writing trials carried out in a joint industry-academic fashion have shown that making a strict distinction between faults and errors is not practical.
- D-System Monitor and D-App Monitor are designed to monitor the state of the system, and while they can detect errors in the form of deviation of parameters from normal values, they cannot detect the faults that cause this deviation. Linking of D-Script actions without regard for the cause is undesirable from the standpoint of failure response.

Writing diagnostic scripts as actions with D-Script allows the diagnostic process to identify SOFs with a clearer cause (i.e., faults) from among those that are more uncertain (i.e., errors of unknown cause). In this way, failure response can be described using a D-Script suited to the level of detail in the D-Case argument concerning the SOF.[2]

---

[2]Faults and errors can only be identified when the cause has been determined.

8.2.4 D-Case Vocabularies

D-Case descriptions are based on natural language, and for this reason, the definitions and meanings of the terms used are highly important. D-Case has thus been provided with a vocabulary for defining terms independently of descriptions appearing in documents. Different vocabularies are provided (such as in the SVBR standard, Agda, and other defined languages) for defining meanings.

D-Script has been designed such that descriptions thereof appearing in D-Cases can only take the form of SOF symbol names, action names (i.e., D-Script function names), and other information that would not hamper readability. A vocabulary provides definitions for these symbol and action names. Meanings can thus verify whether actions given by scripts are valid in terms of executability.

8.3 D-CASE AND D-SCRIPT DESCRIPTIONS

The fundamental role of a D-Case is to present an argument concerning the normal state of the system under consideration. Therefore, they contain no explicit structures like those of the D-Script pattern, i.e., SOF ⇨ Action.

8.3.1 Monitor and Action Nodes

D-Case adds monitor and action nodes as extensions of the assurance case so that the system can be monitored and actions can be executed. The D-Script pattern is seen in these nodes.

- D-Case monitor nodes coordinate with D-System Monitor and D-Application Monitor running on D-RE in order to determine whether or not the operational system is in a normal condition, or in other words, fully within IORs.

- D-Case action nodes describe planned operational procedures, which help to achieve the dependability requirements of the corresponding goal when executed successfully.

D-Case arguments are developed by decomposing dependability-related goals, and for this reason, monitor and action nodes will not appear under the same sub-goal. Furthermore, the action to be taken in the event of an anomalous monitored value will depend on the cause thereof, meaning that monitor and action nodes cannot simply be lined up. In the example shown in Fig. 8-4, a monitor node that calls for monitoring of the state of a hard disk and an action node indicating how to rectify insufficient free space on the hard disk are arranged in different branches of the D-Case. While the relationships

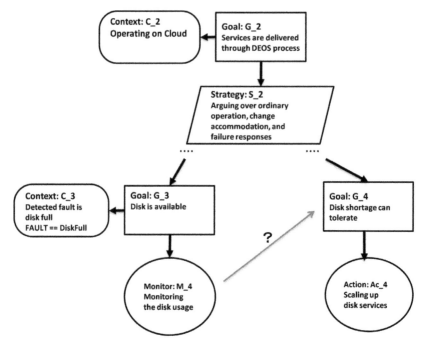

Fig. 8-4 Monitor and Action Nodes.

between the nodes can be easily interpreted by a human in the form of natural language, it would be difficult to implement machine-based processing without making modifications.

8.3.2 D-Script Tags

D-Script tags provide a way to insert the bare minimum amount of information needed for D-Script machine-based processing into natural language D-Case notation without impairing readability. Each tag comprises a name and value, which are delimited by two colons. Figure 8-5 shows an example of D-Script tags *AdminName::* and *Action:* within a D-Case. All natural language outside the tags is ignored for the purpose of D-Script interpretation, meaning that they can be easily inserted into conventional D-Cases.

D-Script tags should normally be added to D-Case context nodes [6]. By defining their own D-Script tags, users can employ them as D-Case parameter nodes in developing an argument. For example, the value of an *AdminName::* tag entered into a context node would be referenced by all sub-nodes underneath it. In addition, these tag values can also be referenced from D-Script action functions indicated in *Action::* tags. It is possible, therefore, to

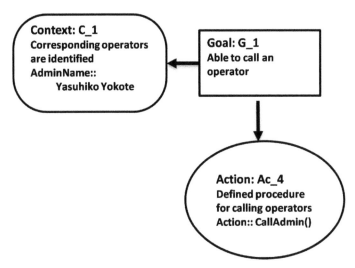

Fig. 8-5 D-Case with D-Script Tags.

customize D-Script actions by changing the D-Script tag values in the D-Case context nodes.

A number of tags with predefined meanings have been reserved for D-Script, and these are shown in Table 8-2. They represent the requisite minimum for creating reliably executable scripts.[3]

Table 8-2 Reserved D-Script Tags.

Reserved tag	Meaning
SignOfFailure::	SOF identified in D-Case Examples: *DiskFull, SystemFailed*
SignOfFailureCase::	Actualized SOF (i.e., observation thereof)
Action::	Function name for action to be executed Examples: *CallAdmin(), RestartServer()*
Location::	Logical location for execution of action Examples: *WebServer, DataStore*
When::	Special timing for execution of action
Presume::	Precondition to be satisfied before execution of action
Range::	Description of IOR

---

[3]The functionality of D-Case monitor and action nodes can be implemented by adding *Action::* and *Range::* tags to the former and an *Action::* tag to the latter.

(1) SignOfFailure: Adding SOF to arguments

Dynamic evidence such as provided by monitor nodes can include SOFs that ultimately lead to system failure. With D-Scripts, these SOFs can be identified and added as context to evidence nodes, and the *SignOfFailure::* tag is used for this purpose. Figure 8-6 shows an example of this type of tag, and multiple SOFs can be identified in this way.

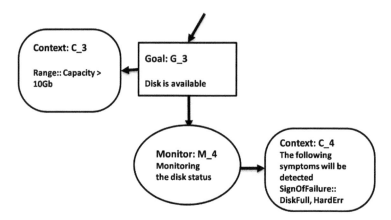

Fig. 8-6 Identification of SOFs using *SignOfFailure* Tag.

(2) SignOfFailureCase: Action for SOF

SignOfFailureCase:: is an extremely basic tag in terms of the D-Script pattern (i.e., *SOF ⇨ Action*). Figure 8-7 shows an example of how this tag is used to describe the action *UpgradeEDS* to be taken in response to *DiskFull*, an SOF identified in a monitor node. In the context associated with the action node, the *SignOfFailureCase::* tag has the value *DiskFull*. This clarifies that the action in question is to be taken in response to actualization of that SOF, and further, identifies the relationship *DiskFull⇨UpgradeEDS()*.

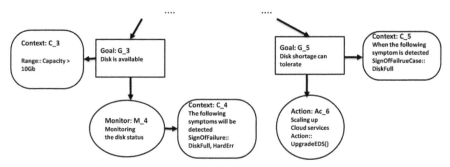

Fig. 8-7 *SignOfFailureCase::* and Action.

It is important that the SOF be already identified within the D-Case using the *SignOfFailure::* tag. This makes it possible to confirm whether or not the SOFs from the D-Case argument are being properly addressed.

It should be noted that, in line with the Open Systems Dependability concept, it is not assumed for D-Script that all causes of failure can be determined in advance. In order to provide for unforeseeable malfunction or anomalies, *SignOfFailureCase::Unidentified* can be used to describe action to be taken in response to unidentified SOFs.

(3) Location: Specifying computer

In consideration of systems comprising multiple computers, D-Script provides a *Location::* tag to indicate which computer should execute specific actions. In order to do so, the tag indicates a logical computer location or type. The *Location::* tag is required in D-Scripts in order to identify clearly where actions are to be executed. Accordingly, actions without this tag will not be executed.

(4) When: Specifying timing

As a mechanism unique to the DEOS Process, monitor nodes are used to monitor the state of the operational system. In practice, however, it is very difficult to detect reliably actualized SOFs through monitoring. Consider, for example, the case of memory leakage—a quintessential software failure. It is practically impossible to determine whether a large amount of data has simply been accessed from available memory and the condition is normal or whether a leak has actually occurred. In order to provide for this type of situation, preventive action based on operation-related experience can be taken as part of regular maintenance to prevent failure from occurring. In terms of our example, the computer in question could be periodically rebooted during regular maintenance before any performance problems due to memory leakage could materialize.

To complement the *SignOfFailureCase::* tag, D-Script also provides a *When::* tag that can be used to describe the execution timing of actions. This tag's value can also be set to "now" if the action is to be executed immediately upon loading of the script.[4] Figure 8-8 shows a *When::* tag that calls for restarting of an application on a regular basis.

(5) Presume: Specifying order

Meanwhile, the *Presume::* tag is used to describe limiting conditions before and after action processing (Fig. 8-9). Whereas the *SignOfFailureCase::* tag describes

[4]In order to comply with the D-Script pattern, the *SignOfFailureCase::* tag is always used to indicate the type of SOF foreseen for the action, even in the case of actions executed based on operation time. In this way, the writing of actions not relevant to the SOF can be avoided.

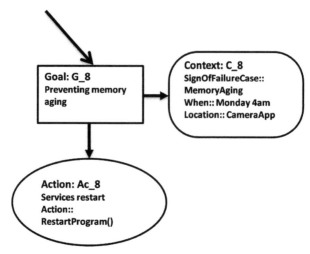

Fig. 8-8 Example of Action in Regular Maintenance.

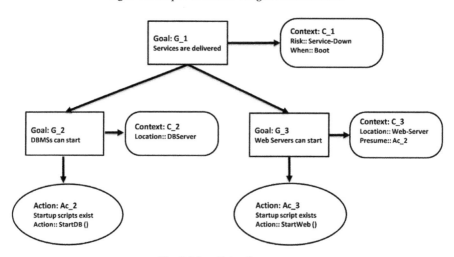

Fig. 8-9 Specifying Sequence.

the assumed execution sequence when an SOF has actualized, the *Presume::* tag describes the assumed execution sequence for successful execution of an action.

In the case of distributed systems comprising a number of computers, *Location::* and *Presume::* tags can be combined to describe operational procedures.

(6) Range: **Specifying in-operation range**

A D-Case is developed as a goal concerning a dependability requirement, with evidence nodes containing test results and the like added to support that goal.

To this argument, D-Scripts add actions—that is, planned steps to be taken at some point in the future. Yet the success of actions cannot be guaranteed. Accordingly, D-Case arguments must be developed such that the addition of actions does not actually reduce assuredness. We have set forth the following minimum requirements for this reason:

- Actions executed by D-Scripts must record and report their success or failure as evidence. In the case of failure, the corresponding execution environment must be logged, and the SOF identified in the D-Case must also be indicated.

- Recovery action to be taken in the event that the D-Case action ends in failure must also be described.

D-Scripts must be written so as to satisfy the above two requirements. D-Script Engine records the execution results of all actions executed from *Action::* tags. *Range::* tags can be added to D-Case monitor nodes to indicate which parameters should be recorded with their In-Operation Range.

8.4 ASSURENOTE—AN INTEGRATION TOOL FOR D-CASE AND D-SCRIPT

AssureNote is a support tool that integrates D-Case and D-Script, and it can be used in design, operation, and all other stages of the system's lifecycle—from D-Case consensus building to script generation, application of scripts to the operational system, and collection of script test results. Needing no special software to be installed, this tool allows D-Scripts to be written and deployed simply using a web browser.

AssureNote is compatible with all other tools developed as part of the DEOS Project, such as D-Case Editor, D-ADD, and the reference D-RE implementation; furthermore, it has been designed to support linking with standard monitoring systems such as Zabbix and Amazon Cloud Watch. It is available as an open-source product from www.assurenote.org.

8.4.1 D-Case Authoring

AssureNote supports basic D-Case authoring based on stakeholder management, which assigns stakeholder classifications to all users and records this information with newly added notation. With D-Cases developed using AssureNote, therefore, it is possible to identify when a description was added, by whom, and in what stakeholder capacity. Typical screenshots of D-Case editing using a pattern library are shown in Fig. 8-10. And as shown in Fig. 8-11, this tool also provides logging functionality to record development of the D-Case over time. The condition of the D-Case at any point in time can thus be easily inspected.

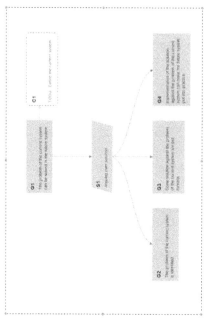

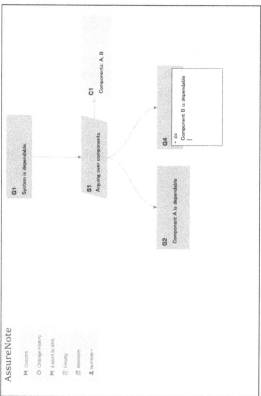

Fig. 8-10 AssureNote D-Case Editing and Pattern Library.

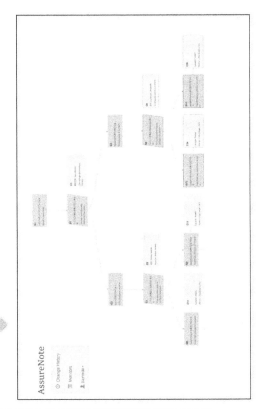

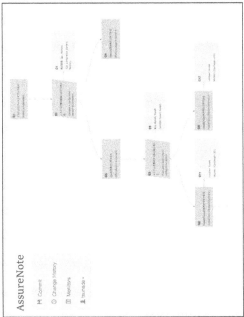

Fig. 8-11 Recording Changes in D-Case over Time.

8.4.2 Defining D-Script Action Parameters

D-Script actions are described in the D-Case with a name indicating the SOF they address. AssureNote can now define the actual action parameters using the D-Shell statically typed shell language.

```
DFaultCallAdmin() {
  if(!(mail -s "urgent" AdminName< ErrorLog.txt)) {
   return UnavailableAdmin;
  }
  return Nothing;
}
```

Fig. 8-12 Typical Action Definition in D-Shell Language.

Figure 8-12 shows an example of an action function in this language. In specific terms, each action is defined as a function with the same name as the corresponding D-Script *Action* tag. Return values from D-Script action parameters are of the DFault type. That is to say, DFault is defined as a type for returning information on the error or fault that caused the failure in the event that execution of a D-Script function ends in failure. If execution does not end in failure, "Nothing" is returned.

The D-Shell language has been designed as the world's first statically typed shell language. For example, the parameter *AdminName* (a global variable) used above must be defined as a parameter in the D-Script tag from the D-Case context node. If *AdminName* were not defined in this way, a type error would result. Static type-checking functionality allows AssureNote to check consistency with D-Script tags and report any coding mistakes.

Based on the relationships between parameters defined using the D-Shell language and D-Script tags (such as *SignOfFailureCase::* and *Action::*), AssureNote can create scripts to be run on D-RE. Furthermore, it can send these scripts to D-RE and execute them. Figure 8-13 shows an example of a

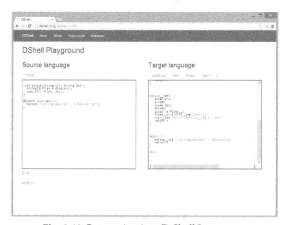

Fig. 8-13 Conversion into D-Shell Language.

script converted into the D-Shell language from a D-Case and D-Scripts. This script would be executable using the D-Shell engine (described below), which is distributed with D-RE.

8.4.3 D-Shell—A Dependable Script Engine

A highly reliable shell-script engine, D-Shell was developed for use with the D-RE reference implementation. This reference implementation and the Linux operating system upon which it is based place considerable importance on easy implementation of command and file operations—one typical example being ready access to internal operating system data using the */proc* filesystem. The D-Shell language has been developed as a more compact version of the Konoha [7] statically typed scripting language, and it makes command and file operations easier to program. It also provides an exception handling mechanism not found in conventional shell languages and can thus handle shell runtime errors.

D-Shell is provided as open-source software together with the reference D-RE implementation. We also have plans to make sample notation for failure response actions using D-RE and Amazon Web Services freely available.

8.4.4 Using Existing Script Engines

Practically all modern operational systems make use of operation scripts. Meanwhile, a wide variety of languages such as the Bourne shell, csh, the Korn shell, Perl, and Python have been developed for scripting language engines and are put to use in operational systems. However, environments combining operation scripts written in multiple languages are far from ideal, and stability issues can also arise depending on the scripting engines used. It is often the case, therefore, that dependability requirements call for a limit on the number of languages used for operation scripts. D-Shell cannot easily be applied in such an environment.

AssureNote provides a multi-script converter (Fig. 8-14) for converting the D-Shell language—the native language for D-Scripts—into the Bourne shell, Perl, or compiled C binary code so that D-Script from D-Case arguments can be widely applied without being tied to a specific script-execution environment. While the converted scripts certainly cannot offer the same level of exception processing and execution log consistency as the D-Shell engine, they make it possible for operation scripts based on the DEOS Process to be put to use in a wider range of operational systems.

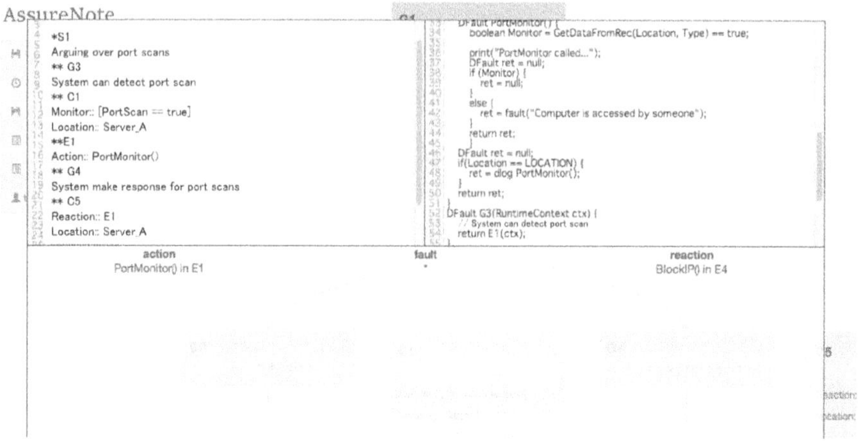

Fig. 8-14 Multi-Script Converter.

8.4.5 D-Case Monitor Nodes & Operation Support

Figure 8-15 shows an example of how AssureNote works with D-Case monitor nodes and a D-Box so that script execution results can be confirmed. Monitor nodes being used to gather evidence are displayed, and past logs can also be viewed as evidence.

8.5 CASE STUDY: ASPEN e-LEARNING SYSTEM

Aspen is an e-learning system for programming that is operated by Yokohama National University (YNU). It was designed and developed based on the DEOS Process by the joint industry-academia Aspen Development Team and is currently used in lectures at Waseda University, YNU, and other higher education institutions. With failure response, stakeholders' agreement, change accommodation, and various other processes implemented and in operation since 2012, Aspen has proved to be a very interesting case study in terms of the DEOS Process. The following sections briefly introduce this system, starting with its D-Case argument for dependability requirements and focusing particularly on how the necessary D-Scripts were added and applied in practice.

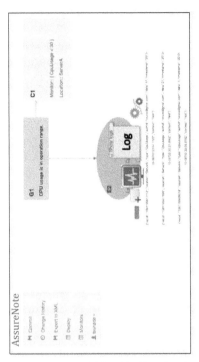

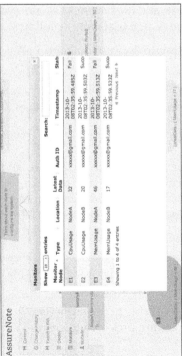

Fig. 8-15 Monitoring Image and Monitor Node in AssureNote.

8.5.1 System and Service Overview

As mentioned above, Aspen is an e-learning system for programming. It has a standard web-based configuration that allows users to access programming assignments, work through these assignments, and submit their work, all using a web browser. Figure 8-16 shows the physical makeup of the system.

Some notable milestones in its operational history are as follows: In 2012, Aspen 1 was planned and first went operational in the spring term at YNU, and Waseda University decided to use the system at the last moment; in June, 2012, failure occurred and countermeasures were applied; and in 2013, the system was transitioned to Aspen2, including Change Accommodation AWS and open source implementation.

8.5.2 DEOS Process & D-Case Development

The Aspen project was conducted in the form of a joint industry-academia partnership as an evaluation trial of D-Case and D-Script notation. While the users of the Aspen system are actual students taking programming courses, experts and consultants from private-sector firms joined the development and operation teams in order to study the D-Case/D-Script description. The DEOS Process Decomposition pattern was employed in the Aspen trial, and AssureNote was also used. The arguments presented by the individual stakeholders were recorded for each sub-process, and the system's D-Case was developed. The stakeholders identified for the Aspen trial were as follows:

- Owner   The Aspen service (i.e., ultimate decision maker for all aspects of service)
- Developers   Those involved in development and testing
- Operators   Those involved in failure response and other aspects of system operation
- Users   University students (i.e., those granted Aspen access)

The first step was to set "Aspen is dependable" as the top goal, and preconditions (i.e., programming class for beginners; half-year duration; 70 participants) were clearly defined as contexts and recorded. The argument was developed by decomposing the goal in line with the phases of the DEOS Process—i.e., the *Ordinary Operation state*, the *Failure Response Cycle*, and the *Change Accommodation Cycle*.

The owner added dependability requirements for the *Ordinary Operation state*, and the following four requirements were added to the D-Case:

- *Availability* The service will always be available to the users (i.e., students). Sufficient resources are provided to ensure accessibility.

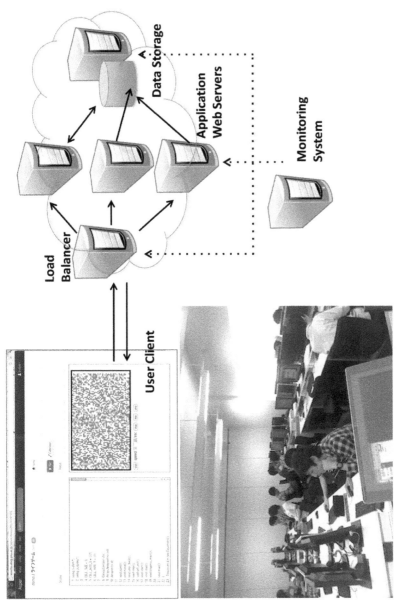

Fig. 8-16 Aspen System Configuration.

- *Reliability*   No hardware or software failures occur during provision of the service.
- *Integrity*   Programming assignments supplied by the owner do not disappear.
- *Privacy*   Personal information is not disclosed to unauthorized parties.

8.5.3 Dependability Requirements and Accountability

Adding test results and other evidence, the developers further expanded the argument in order to satisfy the dependability requirements set by the owner. The next step was to have the developers' D-Case reviewed by the operators and users from their individual perspectives, which constituted the achievement of accountability on the part of the developers. As a result of this process, evidence was added to address insufficiencies, and inconsistencies in context descriptions were fixed.

It is critically important for D-Scripts that SOFs are identified and that the developers and operators agree in the development phase on how these SOFs should be treated in actual operation. The operators are then required to write D-Scripts to prevent these identified SOFs from developing into a system failure. The SOFs are described using D-Script tags in context nodes corresponding to their evidence nodes.

8.5.4 Adding Operation Scripts & Accountability

To develop the goal "Failure response is possible" concerning the DEOS Process, the operators write failure-response procedures to be carried out for each of the SOFs agreed with the developers. In our example, this goal was decomposed according to the ability to monitor for actualization of SOFs and failure-response actions being made available, and the argument for each was developed.

Actions added using D-Scripts can support dependability requirements in that they provide for countermeasures; nevertheless, the execution of actions can end in failure. This possibility is described as another identified SOF, allowing countermeasures to be described for failure of actions.

Operational procedures containing D-Scripts are reviewed in terms of accountability to the owner and users. The concerns of the users are identified and added to the D-Case in the form of new SOFs, and the operators add the corresponding D-Scripts. This process ends in agreement concerning the provision of services.

AssureNote can be used to elicit the correspondence relationships between SOFs and actions from the D-Case notation and to create a list thereof.

In addition, it can also display a warning for any SOF for which no such correspondence has been defined.

8.5.5 Failure and Failure Response

For the purpose of the Aspen Project, D-Scripts were written as a means of mitigating SOFs concerning evidence that was related to the dependability requirements during actual operation. Over the course of multiple reviews, it was confirmed that these scripts were more comprehensive than those that would normally have been written independently by operators; furthermore, there was also a much higher level of consistency in SOF correspondence. That having been said, completeness cannot be guaranteed.

During the Aspen project, we encountered a serious system failure that resulted in programmer e-learning being suspended. The cause of system failure was very basic: the server was unable to deal with an increase in traffic. The D-Case did identify this SOF, and a D-Script had been written to scale horizontally by adding servers. Yet the Aspen assumptions described in the corresponding D-Case context node did not reflect actual operational conditions. It had initially been assumed that students would use the service from home, but in practice, it was also used by students in practical training rooms, causing traffic concentration to exceed expectations. The deployed monitoring system did not observe for increases in traffic, meaning that the horizontal scaling action could not be executed. In addition, when working to address this failure, we identified locations where horizontal scaling would not sufficiently balance the traffic.

Ultimately, a vertical scaling action was added to increase server performance during practice hours only and the D-Case and D-Scripts were updated accordingly. This made it possible to achieve accountability to the users and for consensus to be reached by all stakeholders.

8.5.6 Change Accommodation

The purpose of the DEOS Process is to increase dependability over the full lifecycle of a system by accommodating change in its constantly evolving environment and objectives. In this, operation process logs provide extremely important information for change accommodation.

In the specific case of the Aspen Project, once consensus had been reached by stakeholders (i.e., the developers and the owner after the end of development; the users and owner after the end of failure response), requirements for the system's next embodiment were recorded as change accommodation goals. Operation logs served as one source of evidence for determining whether or not these requirements were appropriate. When D-Scripts are used to control operation, logs are retained for all actions. We made use of this information in order to develop the argument.

Based on the D-Case argument, the Aspen Project implemented change accommodation as follows to progress from Aspen1 to Aspen2:

- Reduced dependence on software developed in-house and adopted the open-source solution Moodle.
- Transitioned to Amazon Web Services (the cloud computing platform) as an alternative to self-operated servers.

Furthermore, two new dependability requirements were added for Aspen2 based on the change accommodation argument, and the system was reconfigured and operated accordingly. The D-Case after change accommodation grew to include over 200 nodes.

8.5.7 Conclusions from Aspen Trial

We have made the D-Case and D-Scripts developed for the purpose of the Aspen trial freely available at www.assurenote.org. Before finishing this discussion, let's take a look at some statistical information on growth of the D-Case and D-Scripts during the course of this trial. Figure 8-17 shows how the numbers of the following items grew during that period:

- Dependability requirements
- D-Case goal nodes
- D-Case context nodes
- D-Case evidence nodes

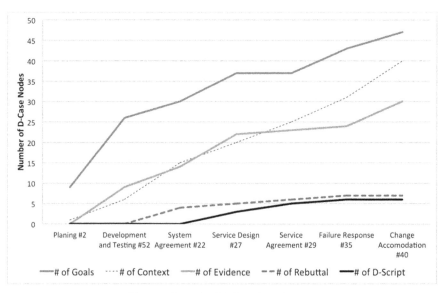

Fig. 8-17 Growth of D-Case & D-Scripts.

- SOFs identified during consensus building
- D-Scripts to address SOFs

Dependability requirements represented the starting point of the Aspen trial, and the number thereof stayed the same until change accommodation. Multiple rounds of stakeholders' agreement in line with the DEOS Process led to an increase in the numbers of goal nodes, context nodes, and identified SOFs. This resulted in an increase in the number of operation scripts. However, D-Script actions have not been added for all SOFs, and other approaches such as the elimination of SOFs before execution are being employed.

The Aspen trial allowed us to verify that operation scripts could be derived from dependability requirements based on stakeholders' agreements, and in addition, that multiple rounds of stakeholders' agreement in line with the DEOS Process allowed for a large number of SOFs to be addressed.

8.6 CURRENT SITUATION

D-Scripts provide a means of executing control procedures developed from a D-Case dependability argument and based on stakeholders' agreement in an operational system without the need for software redevelopment. This makes it possible to assure system dependability in a reliable manner from design right through to operation and to achieve consistent accountability.

An open-source version of AssureNote can be obtained from www.assurenote.org. Anybody with an interest in D-Cases and D-Scripts can freely use this tool to develop samples. Meanwhile, licenses for D-ADD and other commercial software can also be acquired from this website in order to integrate them into actual systems.

D-Shell, our reference D-Script Engine implementation, is available as part of the overall reference D-RE implementation. In addition, we are currently preparing to release D-Script specifications in the form of a rule set to accompany the D-Case specifications. Sample scripts developed for D-Case and D-Script testing using D-RE, Amazon Web Services, and HPC cluster systems are also available.

REFERENCES

[1] Kelly, T. and R. Weaver. 2004. The Goal Structuring Notation A Safety Argument Notation, IEEE/IFIP International Conference on Dependable Systems and Networks (DSN 2004).
[2] Bloomfield, R. and P. Bishop. 2010. Safety and Assurance Cases: Past, Present and Possible Future—an Adelard Perspective, in Making Systems Safer. Proceedings of the Eighteenth Safety-Critical Systems Symposium, pp. 51–67.
[3] Tokoro, M. (ed.). 2012. Open Systems Dependability: Dependability Engineering for Ever-Changing Systems, CRC Press.
[4] Avizienis, A., J.-C. Laprie, B. Randell and C. Landwehr. 2004. Basic concepts and taxonomy of dependable and secure computing, IEEE Trans. Dependable Sec. Comput., Vol. 1, No. 1, pp. 11–33.

[5] Kinoshita, Y. and M. Takeyama. 2013. Assurance Case as a Proof in a Theory: towards Formulation of Rebuttals, Proceedings of the 21st Safety-critical Systems Symposium, SCSC.

[6] Matsuno, Y. and K. Taguchi. 2011. Parameterized Argument Structure for GSN Patterns, Proceedings of the 2011 11th International Conference on Quality Software, IEEE Computer Society.

[7] Kuramitsu, K. 2011. Konoha Script: static scripting for practical use. In Proceedings of the ACM international conference companion on Object oriented programming systems languages and applications companion, SPLASH '11, pp. 27–28, New York, NY, USA, ACM.

9

D-ADD—THE AGREEMENT
DESCRIPTION DATABASE

This chapter takes a detailed look at the Agreement Description Database (D-ADD), which not only provides functionality for managing the history of D-Cases but also processes, records, and retains all of the information related to stakeholders' agreement throughout the lifecycle of a system in order to support DEOS Process accountability. D-ADD can be used to achieve the following:

(1) Connecting the D-Case and the system and achieving rapid response in the form of, for example, failure prediction and avoidance.

(2) Real-time checking of consistency between evidence contained in the D-Case and actual data from monitoring of the system as well as rapid recovery in the event of divergence.

(3) Recording of data pertaining to all failures—including those that could be avoided—in order to prevent the same or similar failure from occurring again in the future.

This chapter describes the workings of D-ADD with a particular focus on how to achieve the above. We start by looking at the architecture of D-ADD as it fits into the overall DEOS Process. We then describe how D-ADD is used to support the achievement of accountability in the DEOS Process using a case study involving a commercial broadcast system. After this, we outline an implementation of D-ADD, which is available through our website. We then round off the chapter with a look at how D-ADD could contribute in a number of actual business situations, as well as how this technology could innovate the software development process.

9.1 THE D-ADD ARCHITECTURE AND THE DEOS PROCESS

D-ADD is accessed by all of the sub-processes that make up the overall DEOS Process. In terms of the Consensus Building process, it provides for reliable recording of the argumentation procedure, searching for related arguments and relevant case studies, structuring of arguments, and other related tasks. The arguments recorded in D-ADD must be accompanied by various kinds of evidence. Accordingly, documents on stakeholder requirements, plans, specifications, contracts, and all other related documents, as well as D-Cases themselves, must be stored and linked in this database. In cases where a system consists of multiple subsystems with different ownerships, the corresponding D-ADDs must provide functionality for exchanging the necessary information with one another.

In the Development process, D-ADD supports accurate implementation of the agreements from the preceding Consensus Building process. Documents, programs, code, and other material developed or used in this process are stored in the repository discussed previously. In the Failure Response process, D-ADD is used to determine the cause of the failure and to provide important information pertaining to the required countermeasures, avoidance, and so forth.

D-ADD supports the Accountability Achievement process by providing information required in order to analyze causes of failures, make repairs, execute countermeasures, and launch new services. For this, D-ADD retains all the operation logs. In the *Ordinary Operation state*, it interacts with the operational system to confirm whether OSD is assured, and if necessary, to initiate the *Failure Response Cycle* or the *Change Accommodation Cycle*. In this way, D-ADD plays a vital role in implementing the DEOS Process.

9.1.1 D-ADD Architecture

This subsection describes the architecture of D-ADD in terms of three elements—namely, (1) Fundamental Tools, which provide interfaces and functions to the above-mentioned processes and state, (2) a Core, which implements the models envisaged and manipulated using the tools, and (3) a Hybrid Database, which is composed of heterogeneous databases and retains the models. Figure 9-1 illustrates the overall architecture of D-ADD.

(1) Fundamental tools

Three tools have been provided as a fundamental means for interacting with D-ADD:

 a Consensus Building Support tool, an Accountability Achievement Support tool, and a Monitoring tool.

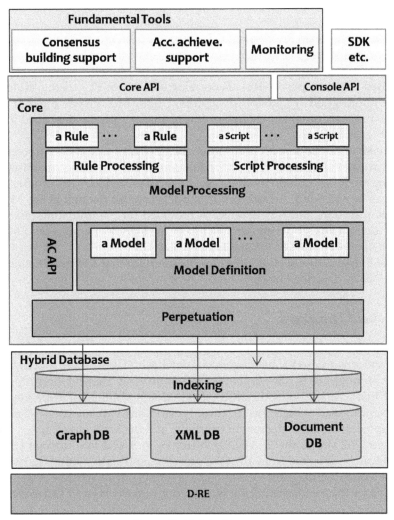

Fig. 9-1 D-ADD Architecture.

As its name suggests, the Consensus Building Support tool is intended for use within the Consensus Building process in order to ensure that the process proceeds smoothly. The Accountability Achievement Support tool, meanwhile, has been developed for use within the Accountability Achievement and Failure Response processes. It provides support for accountability to stakeholders on the basis of information stored in D-ADD, not only upon the launch of new services, but also when failure avoidance processes are executed in the information system or when a failure has actually occurred in that system. Meanwhile, the Monitoring tool would typically be used within the *Ordinary Operation state* to support monitoring of the operational system in conjunction with D-Case monitor nodes.

(2) The Core

The above-mentioned Fundamental Tools access the Core using the Core API. The Core consists of three layered components—namely, the Model Processing, Model Definition, and Perpetuation sections. The Model Processing section is divided into two parts—Rule Processing and Script Processing. Rules to change information in a model are described in the Rule Processing part. The actions that carry out these changes in the form of scripts are described in the Script Processing part.

The Model Definition section defines the model for each type of document to be processed. Such models include a Fundamental Data model, a Consensus Building model, a Toulmin model, a Meeting model, and a D-Case model, and these can be extended. Actions on the models are also defined in the Model Definition section, and functionality is provided for manipulating relations among documents, between agreements and stakeholders, between D-Case and the operation state of the system, and so forth.

The Perpetuation section provides the interface to the Hybrid Database, which will be described below.

(3) Hybrid Database

The Hybrid Database retains all of the information on the models described above and is composed of heterogeneous databases. Considering that the type of data in question is non-structured in nature, that inter-relationships are critical, that time attributes are added, and that large volumes of disparate data must be processed at high speeds, three types of database are used—namely, the graph database, the document database, and the key-value store.

Figure 9-2 shows the D-ADD architecture in ArchiMate® format [1]. As D-ADD can be accessed from all of the processes and states within the overall DEOS Process, this diagram focuses on the interfaces (shown by the –o symbol) provided by the Fundamental Tools, the Core, and the Hybrid Database.

9.2 ACTUAL USE OF D-ADD

9.2.1 D-ADD and Accountability

The DEOS Process requires that accountability be achieved whenever a failure occurs in the operational system. For example, the provider of a service may need to be accountable to its users and other stakeholders by describing the failure itself, the cause, damage suffered as a result, and plans for recovery. When providing this type of information to stakeholders either inside or outside the company, it may be necessary, for example, in the case of a serious failure with major ramifications for society at large, for the company president, a director, or another representative to hold a press conference. In such a case,

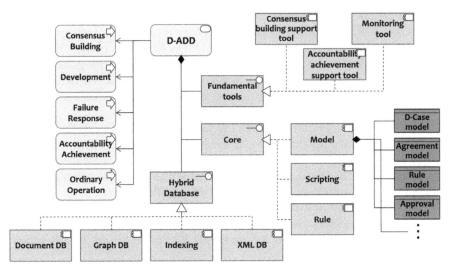

Fig. 9-2 D-ADD Architecture in ArchiMate® Format.

the company's public relations department would likely prepare explanatory documents. If, on the other hand, the failure was of a minor nature—perhaps general users may have been unable to connect to the Internet for five minutes or thereabouts—it may suffice for an apology to be posted on a web page catering for service users. We can see, therefore, that the method, timing, and content of explanation for the purpose of accountability depend on the magnitude and nature of the failure, the type of service affected, and the parties to whom the company must be accountable.

Although D-ADD houses the information that stakeholders would need in order to achieve accountability, it provides no functionality for the preparation of, for example, press conference material by a public relations department. From the perspective of accountability, the role of D-ADD is to manage data in such a way that, whenever called on, it can provide information on the current state of failure response as well as evidence that backs this up. In parallel with the *Failure Response Cycle*, D-ADD support for accountability starts with failure prevention and includes responsive action and cause analysis; furthermore, if deemed necessary in these phases, the initiation of a *Change Accommodation Cycle*, the building of new consensus among the stakeholders, and the release of improvements (including functionality for preventing reoccurrence of the failure in question) are also supported. Below, we summarize the accountability achievement sequence managed using D-ADD.

D-ADD constantly records operation logs. When a failure occurs, the corresponding D-Scripts are executed automatically to initiate responsive action. If this cannot be achieved automatically, pertinent failure information is recorded in D-ADD and operators implement rapid response in a manual fashion. In the Cause Analysis phase, evidence information stored in D-ADD (i.e., fundamental documents, agreement logs, and operation logs associated with the D-Case's evidence and context nodes) is examined on the basis of the D-Case and the fault that caused the failure is identified. Logs generated during this process are also stored in D-ADD as further evidence. Based on the fault identified as the cause, the stakeholders involved in this phase debate actions required to prevent reoccurrence until consensus is reached. At this point, the *Failure Response Cycle* ends, and the DEOS Process transitions to the *Change Accommodation Cycle*.

The stakeholders enter a phase of requirement elicitation on the basis of the failure response actions. These actions can only address the confirmed cause; however, in this Requirement Elicitation & Risk Analysis phase, it can be expected that the actions would ideally be enhanced in line with the DEOS Process objective of improving the information system in an iterative fashion. Completely new functions could, therefore, be debated and agreed upon. At the end of this phase, the D-Case is updated to reflect the necessary failure response and other new requirements. The information system is then modified, tested, and released.

At this stage, one preparatory step for achieving accountability has been completed. We designed D-ADD to function as a repository for tracing—even after a certain period of time has passed—revisions and new functionality added in the *Change Accommodation Cycle* to actually address the corresponding failure and expected improvements.

9.2.2 Using D-ADD—A Commercial Broadcast System Case Study

A commercial broadcast system is an essential, intrinsic part of any commercial broadcaster. In contrast to public broadcasters, these companies obtain most of their revenue from the broadcasting of TV commercials based on contracts with sponsors. For this reason, information technology was rapidly adopted in the management of related sales activities and broadcasting schedules [2] soon after these broadcasters were established. Since then, commercial broadcast systems have been enhanced with functionality for managing TV-show and data broadcasting content, the preparation of cue sheets, and so forth, thereby evolving into massive, highly complex systems catering for multiple departments and many individual users. Furthermore, these systems also ensure that sponsors are invoiced for the commercials that have been broadcast. Commercial broadcast systems may be purchased as over-the-counter packages [3, 4], but as these solutions must then be customized to suit the needs of the particular broadcaster, they too tend to become highly specialized and complex.

Commercial broadcasters must be licensed and their activities are required to comply with the Radio Law and the Broadcast Law [2]. Recent years have seen a spate of accidents causing broadcasts to be halted, overbilling based on untruthful numbers of TV-commercial airings, and other scandals involving similar falsification. In response, the authorities have issued a directive instructing broadcasters to prevent reoccurrence and make this type of incident a thing of the past [5]. Any commercial broadcaster that seriously fails to comply could even have its license revoked; accordingly, it is now more important than ever that dependability be assured, not just for broadcasting equipment, but also for the commercial broadcast systems themselves.

As described in previous chapters, the DEOS Process has been specifically developed to assure OSD over the complete lifecycle of massive, complex systems that must accommodate change, and as such, it is ideal for application to commercial broadcast systems. This process is based on the principle that system dependability can be enhanced by dual-loop operation comprising the *Failure Response Cycle* and the *Change Accommodation Cycle*. In order to make this possible, a D-Case must be developed and the stakeholders must agree on the arguments that it makes so that accountability will be achievable. D-ADD provides support for these activities.

(1) Issues in consensus building in a commercial broadcast system

Commercial broadcast systems are usually extremely large, involve multiple departments, and cater for a great many users. In specific terms:

- The editing department plans and manages programming makeup and schedules;
- The sales department seeks sponsors, makes contracts, and manages advertising content for broadcast; and
- The broadcasting department manages program content, final broadcast status data, and broadcasting equipment.

The commercial broadcast system must cover an extremely wide range of activities in order that the roles of all three of these departments can be systematized. And because these activities were originally carried out manually by employees, the individual expertise and experience of each employee are reflected in the corresponding functionality. As a result, each function is specialized and refined for the operations of a particular department. In addition, these functions are required to interact in a complex manner, both within specific processes and between multiple processes.

Given all of these factors, the size, complexity, and specialization of the D-Case for a commercial broadcast system are directly proportional to the extent of the system's characteristic features. Normally, the entirety of a system encapsulating specialist knowledge is understood in detail by very few people or perhaps none at all, meaning that no single person can develop its D-Case. In practical terms, furthermore, it is also impossible for the most senior manager to understand the complete D-Case, to take responsibility for it, or to agree on its content. Inevitably, specialist employees from individual departments must be defined as stakeholders, agree on the content of the D-Case, accept responsibility for that content, and be accountable. For these reasons, it is extremely difficult to reach consensus in the case of a massive, complex D-Case. The following list identifies four particular challenges in this regard:

1. Assurance of D-Case consistency;
2. Estimating the extent to which localized corrections will affect the overall D-Case;
3. Facilitating D-Case development by multiple persons; and
4. Clarification of the extent of responsibility of stakeholders.

The following describes a consensus building approach that can be put to effective use in meeting these challenges.

(2) V-model approach to consensus building

The V-model lifecycle can be used to construct massive, complex D-Cases and achieve agreement among stakeholders. Agreement can be phased to best suit the V-model achieved by combining the hierarchical structure of the D-Case with the organizational structure of the company. Using this approach, the authority and responsibility delegation levels of the company are reflected in the D-Case structure, and agreement comprises two separate phases—planning and verification.

(1) Consensus-building arenas

Consensus building for a certain objective is undertaken at meetings by multiple parties with different interests in a particular matter. Borrowing from *Consensus Building* by Takehiro Inohara [6], these circumstances required for consensus building can be referred to as "ba". The concept of "ba" was proposed by Japanese Philosopher Kitaro Nishida [7], further developed by Shimizu [8], and adopted by Nonaka [9] for knowledge creation. A "ba" is a shared space for communication and knowledge creation. It can be interpreted as a set of circumstances, a situation, or a context, but here, we will use the word "arena" to emphasize that it implies a place as its original meaning. Persons with the appropriate responsibility must come together at a consensus-building arena, and the content to be agreed upon must be understood there. Meanwhile, these arenas are closely associated with the organizational structure of the company.

The organizational structure of a broadcaster has a hierarchical format. In this type of company, departments are divided up according to business objectives, and each realizes a different function. Furthermore, these departments are also broken down into smaller groups tasked with different duties, and leaders are assigned at each level. For example, the company head serves as the top-level leader, typically followed by department chiefs, and then group leaders. Each leader is responsible for the work of his or her team, and they assume authority over resources, give instructions, and receive reports from lower-level leaders. Meanwhile, leaders also hold discussions with lower-level team members and others in order to make decisions and to reach consensus among themselves.

Decision-making meetings are attended by persons with the appropriate decision-making authority and responsibility, persons with responsibility for, and authority over, resources needed to execute the decisions under consideration, and persons with expertise concerning the matter to be decided. Consider, for example, a meeting at which a mid-term business plan must be decided upon: the company head and directors would attend, but lower-level employees could not. Meanwhile, in the case of a meeting concerning the user manual for a new product, the developers of the product would attend, but upper levels would not be represented. In this type of organizational structure where authority is delegated in a top-down fashion, the hierarchical levels place restrictions on how consensus-building arenas can be established.

The D-Case also has a hierarchical structure. The top goal describes a dependability attribute affecting the entire information system in abstract terms. Comprehensive context nodes relating to the entire system are added at the top level, and the definitions and conditions defined therein affect the D-Case as a whole. As the goal is decomposed through successive layers of sub-goals, the dependability-related content becomes more concrete and the context nodes get more specific.

We can thus align the hierarchical structures of the D-Case and the company with one another. The D-Case is developed in line with the hierarchical consensus-building stages and arenas stemming from the broadcaster's organizational structure. Figure 9-3 shows an example of how these stages and arenas could be arranged.

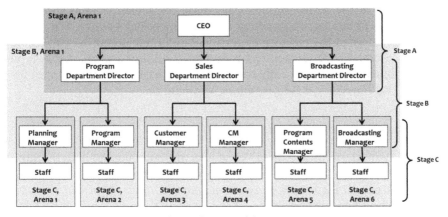

Fig. 9-3 Stages and Arenas.

First of all, three stages (A, B, and C) can be defined based on the broadcaster's organizational structure. Each stage is arranged so as to include two levels of corporate authority and provides for consensus building between these upper and lower levels. Content agreed upon in Stage A is passed down to Stage B as an upper-level decision (as described in the D-Case goal and context nodes). There, negotiation is conducted within the scope of that decision, and the result of consensus building is passed down to Stage C. In other words, instructions relating to the broadcaster's business direction and the delegation of responsibility and authority are effected in a stage-wise manner from the top organizational level through to the bottom. The descriptions in the corresponding goal and context nodes must be consistent when decisions of this nature are passed down to lower levels. If necessary, meanwhile, multiple consensus-building arenas can be established in any particular stage, although each stage must have a minimum of one. The objectives assigned to individual arenas reflect the D-Case notation steps described in Section 4.3, but these can be rearranged or made more specific based on the organizational structure.

Stage A has a single arena. Its objective is to select the D-Case's top goal and to define clearly the context nodes concerning the development and operation of the system as a whole and affecting the related arguments. When consensus building in Stage A ends, it means that responsibility and authority for all subsequent work is delegated to the stakeholders of Stage B. The objective of the Stage B arena is to decide upon a basic configuration for the overall D-Case and to coordinate between the different departments. The arena participants review the decisions made in Stage A (i.e., instructions and scope of authority), clarify the relevant assumptions, and come to an agreement on how the D-Case should be decomposed in Stage C. The objective of the Stage C arenas is to complete the D-Case. To this end, the argument as yet incomplete in Stage B is successively decomposed until each sub-goal can be supported by an evidence or monitor node. Arena participants are selected on the basis of the corresponding objectives, the organizational structure, and the rules of the decision-making meetings mentioned above. The leader must have a role in the relevant department and understand the applicable context. Hierarchically structured consensus-building arenas of this type make it possible to systematically configure and agree upon massive, complex D-Cases. We now look at the procedure for doing so.

(2) V-model for consensus building

Using the V-model approach, the D-Case structure and the consensus building process are first split into a planning agreement phase and a verification agreement phase. Figure 9-4 shows a conceptual diagram of the V-model consensus building process.

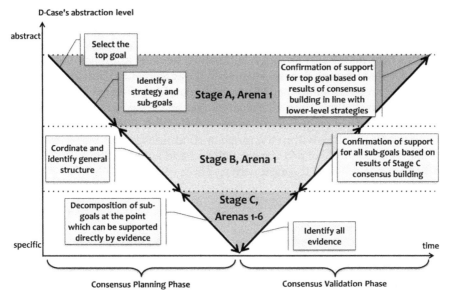

Fig. 9-4 V-Model Diagram.

(a) Planning agreement phase

The objective of the planning agreement phase is to define and agree upon all elements of the D-Case other than its evidence nodes. When using this type of decision-making and authority-delegation model based on the company's top-down structure, the D-Case is developed successively from the top. The arenas illustrated in Fig. 9-3 and the D-Case notation steps from Section 4.3 are applied in order to do so.

Stage A, Arena 1

The objective of Stage A, Arena 1 is to select the top goal and identify context nodes affecting the overall D-Case. These decisions will also include constraints on the information system as a whole, such as operation policies, budgets, and so forth. Because decision making of this nature will affect all lower-level decision making processes, it needs to be undertaken in line with the duties and authority of the upper level of the company's organizational structure. Accordingly, agreement must be achieved in Stage A.

The context nodes linked to the top goal must be applied as context to all lower-level nodes [10]. If it is determined that the scope of any lower node should extend beyond the upper-level context, the argument will need to be traced back to the upper level, where the corresponding conditions must be changed and consensus building redone.

The first consensus-building step in this arena is to propose the top goal (*G1*). Multiple goals may be suggested at this time, and in order to determine which of these is most appropriate, restrictive conditions are described in the corresponding context nodes. The next step is to add a strategy node to describe how the proposed goal will be argued, and sub-goals can be put forward in order to help define this strategy. Ultimately, the best approach for the top goal is decided through debate. Then, with the agreement of the stakeholders, G1 and S1 can be added to the D-Case. Sub-goals identified as G2 through G4 are not finalized at this time. Clarification and finalizing of these sub-goals is the responsibility of Stage B. Further, the stakeholders from the Stage B arena can change G1 and S1 within the range of conditions set for Stage A. Figure 9-5 shows an example of a D-Case developed and agreed upon in Stage A, in addition to the scope of the argument and area of responsibility.

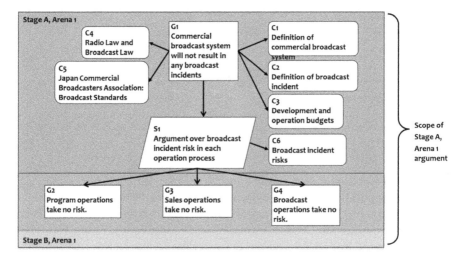

Fig. 9-5 Sample D-Case for Stage A, Arena 1.

Stage B, Arena 1

The objective of Stage B, Arena 1 is to establish an overall plan (corresponding to the D-Case notation steps described in Section 4.3) within the range of restrictions set in Stage A. Given that decisions affecting the overall structure of the D-Case are made at this level, arenas should not be established in each department; instead, a single arena should be established for this stage so that the work of the company as a whole can be interpreted in abstract terms. Stage B stakeholders, who have an understanding of how work is actually performed, must accurately relay to Stage C the approach to G1 and S1 as presented by the Stage A stakeholders, as well as their decisions. Further, in order that individual D-Case branches may be developed on a department-specific basis in Stage C, the Stage B stakeholders must resolve in advance any issues affecting multiple departments and present their findings as sub-goals.

Actual consensus building in this arena starts with clarification of G2, G3, and G4—the sub-goals passed down from the Stage A arena. To do so, the stakeholders verify whether or not G2, G3, and G4 can be derived from G1 based on S1 and whether they support G1. When adding conditions to G2, G3, and G4, it is important to confirm that they are within the scope (i.e., context) defined by C1 through C6 and that they are consistent overall. Next, strategies and sub-goals must be identified so that each of G2, G3, and G4 can be independently refined by the Stage C arenas, and the stakeholders must agree on the D-Case finally arrived at. With this top-down model, authority is delegated from the top level through to the bottom. As long as changes are not required for G1 and S1, which are the responsibility of the top-level stage, there should be no need to seek the opinion of that stage's participants when modifying G2, G3, and G4. However, the upper-level stakeholders should be able to confirm at any time how the D-Case under development is being described in notation and debated. The example in Fig. 9-6 represents the D-Case developed and agreed upon in Stage B, Arena 1 and also shows the scope of the argument undertaken in that arena and its area of responsibility.

Stage C, Arenas 1 to 6

The objective of Stage C is, within the range of restrictions set in Stage B, to decompose the sub-goals to the level at which they may be directly supported by evidence. As the D-Case is successively refined in this way, more and more specialist knowledge will be required; accordingly, six individual arenas are established on a departmental basis so that each can reach consensus independently. That said, certain content may need to be debated simultaneously by multiple departments in order to reach consensus.

Consensus building in each of the Stage C arenas starts with clarification of the sub-goals (G6 through G26) passed down from the Stage B arena, and based thereon, verification of whether the upper-level strategies are satisfied and the upper-level goals can be supported. When adding conditions to each goal, it is important to confirm that they are within the scope of the corresponding upper-level context nodes. This refinement of sub-goals and strategies continues in a successive manner until evidence can be directly linked to the sub-goals. Meanwhile, the stakeholders must agree upon the final D-Case derived through this argumentation process. Each Stage C arena develops a branch of the D-Case, and while these will ultimately form part of the information system's overall D-Case, no evidence nodes will have been added at this time.

Even when multiple departments must participate in development of the D-Case, setting up suitable arenas in line with the content to be debated is highly effective in clarifying the scope of each branch of the argument as well as the corresponding areas of responsibility. When boundary lines are clearly defined in this way, it becomes much easier to identify the overall effect of the localized corrections that must be made. The simplified example in Fig. 9-7 represents the D-Case developed and agreed upon in Stage C, Arenas 1 to 6

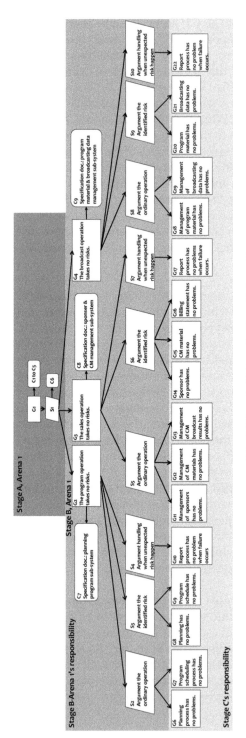

Fig. 9-6 Sample D-Case for Stage B, Arena 1.

from Stage B, and the scope of the argument undertaken in those arenas and their individual areas of responsibility are also shown.[1]

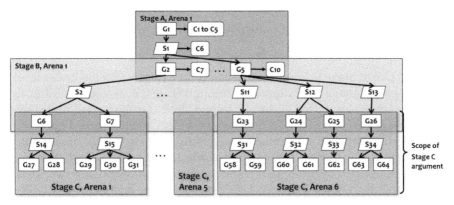

Fig. 9-7 Sample D-Case for Stage C, Arenas 1 to 6.

(b) Verification agreement phase

The objective of the verification agreement phase is to complete the D-Case by adding evidence and to allow all stakeholders to verify that the D-Case is complete. In this phase, verification of the D-Case proceeds in a bottom-up manner in parallel with the organizational structure.

Stage C, Arenas 1 to 6

The objective of Stage C, Arenas 1 to 6 is to add evidence to all leaf goals and to individually reach consensus that the goals in each particular area of responsibility have been supported. Consensus building in each of these arenas starts with the arena participants examining evidence for each leaf goal in order to determine whether or not it fully supports the goal. If deemed adequate, the evidence is linked to the leaf node in the form of an evidence node. In the case of dynamic evidence obtained by monitoring the operational system, D-Script notation is added as described in Chapter 8 in order to implement monitoring and alerts.

Following this, it is determined whether or not sub-goals now supported by evidence fully validate the approach adopted for the strategy node immediately above. D-Case branches individually verified in this way (for example, S14, G27, and G28 from Fig. 9-7) serve as evidence for their upper-level goals (i.e., G6). Once this D-Case branch is confirmed to be free of

---

[1]In order that the D-Case branches developed individually in each of the Stage C arenas can be added to the overall D-Case, they must be checked for consistency. However, consensus building methods within specific layers are exempted from this checking in order to focus on inconsistencies between different layers.

problems, the participants of Arena 1 verify whether *G*6 is supported. When each of the Stage C arenas determines in this way that the D-Case branch decomposed into its area of responsibility is valid through repetitive bottom-up verification of support for leaf goals and strategy confirmation, consensus can finally be reached on that particular D-Case branch.

Stage B, Arena 1

The objective of Stage B, Area 1 is, based on the premise that the lower-level D-Case branches have been supported through Stage C consensus building, to verify that the goals within its area of responsibility are supported and reach agreement in this regard. In addition, content independently agreed upon in a parallel fashion by each of the Stage C arenas must be checked for consistency. As in Stage C, this process starts with bottom-up verification of whether the goals are supported by the evidence and whether the strategies are validated. The fact that the D-Case has been validated within the area of responsibility of the lower-level Stage C serves as evidence to support the Stage B goals. The Stage B stakeholders verify whether the goals in their area of responsibility are supported based on this evidence, and if no problems are identified, final consensus is reached for this stage. If, however, it is determined that the goals are not supported, the relevant Stage C arenas must be requested to repeat the process of consensus building for the corresponding section of the D-Case.

Stage A, Arena 1

The objective of Stage A, Arena 1 is to confirm that the top goal has been supported. As in Stage B, this involves verification that the lower-level branches have each been supported, and on the basis thereof, it can be agreed that the entire D-Case is complete.

In summary, the consensus-building method for huge, complex D-Cases as described above resolves the issues identified above by a) setting consensus-building stages and arenas based on the broadcaster's organizational structure, and b) aligning the degree of D-Case content abstraction with the level of responsibility and authority assigned to each consensus-building arena. By applying the organizational structure to consensus building, upper-level decisions, instructions, and authority can be transferred as the results of top-down consensus building in the planning agreement phase, while the verification agreement phase must confirm that this transfer from upper to lower levels has been properly implemented. Confirmation of this fact is relayed using D-Case goal and context nodes. Normally, confirmation focuses on goals, but careful consideration must also be given to the inheritance of context.

(3) D-ADD support for consensus building

We have provided four application tools to support the consensus building method described above, thereby solving the four issues identified in (1).

(1) Assurance of D-Case consistency

As one of D-ADD's support functions for D-Case consistency, the dictionary function can be used to ensure terminology consistency when developing D-Cases. In order to do so, it manages dictionaries both for individual users and for organizations. Each D-ADD user can define terms related to the system under consideration. Terms that are put to common use across an organization can be systematically managed in the form of organization dictionaries. Users can then confirm that the descriptive text they enter is true to the meanings defined in the dictionaries, thereby ensuring consistency in usage.

D-Case in Agda [11], which was described in Chapter 6, is a proof support language that can be used to express the logical meaning of terms. Registering Agda expressions in line with dictionary vocabulary meanings allows D-ADD to convert the D-Case into formal Agda-based proof. Furthermore, the Agda proof assistant can also be used to check correctness.

(2) Estimating the extent to which localized corrections will affect the overall D-Case

Whenever the information system must be modified in response to changes in specifications, applications, limiting conditions, and the like, the D-Case must also be changed accordingly. In the case of a massive D-Case, however, modification of just one node can have a wide-ranging effect that is very difficult to estimate. As a solution, D-ADD facilitates two distinct mechanisms that support D-Case modification—namely, locating specific vocabulary using the above-mentioned dictionary function and narrowing down the potential range of influence based on organizations' areas of responsibility. To this end, the D-ADD indexing function can be used to identify D-Case nodes containing vocabulary defined in a dictionary.

D-ADD can also manage information on organizations, employees, and authorizations. And within the database, D-Case branches or nodes are associated with this organizational information so that areas of responsibility can be clearly defined. In this way, D-ADD can indicate the organization with responsibility for the D-Case section that must be modified and the full extent of that responsibility. This allows the user to identify boundaries between different areas of responsibility. When making localized changes to a D-Case that affects multiple departments, the D-ADD user can thus utilize both of these solutions to estimate the extent to which the overall D-Case will be affected.

Similar to the above, in situations where D-Case in Agda can be used, a more machine-based approach can be applied in order to identify the overall effect of a planned correction. When the correction has been made, D-ADD can again check the entire D-Case, and any sections that do not pass this check will correspond to the area affected by that change.

(3) Facilitating D-Case development by multiple persons

In situations where more than one person is involved in the development of D-Case notation, D-ADD can provide support by linking with D-Case editing tools such as D-Case Weaver and D-Case Editor as introduced in Chapter 5. In 2) above, we described how a repetitive process is employed to develop the D-Case and reach consensus as debate takes place in different arenas. In this regard, we assumed that the persons responsible for D-Case notation in each arena would actually write D-Case description reflecting development of the argument in plain sight of the stakeholders. Multiple proposals would likely be put forward during the course of debate. In such a case, the Alternative Selection Decomposition pattern identified in Section 4.6 could be used to prepare the D-Case description, with the selection ultimately made during final consensus building. Branches or nodes rejected at this time are nevertheless retained as evidence in D-ADD, providing valuable information for the Accountability Achievement process—a critical part of the overall DEOS Process.

Meanwhile, when refining the D-Case, it is crucial that the argument can be developed in parallel in multiple arenas. As a database, D-ADD provides support for these multiple, simultaneous operations on a single D-Case. In doing so, it retains the complete history of a D-Case developed in this way, and any user with access rights can review the latest version at any time.

(4) Clarification of the extent of responsibility of stakeholders

With functionality for identifying associations between the company's departments and employees and the D-Case branches and nodes, D-ADD provides support for clarification of the extent of stakeholder responsibility. In the event that responsibility for a particular branch or node has not been clearly defined, D-ADD will issue a warning to users. In addition, D-ADD makes it possible for all stakeholders to determine the status of any D-Case section in terms of whether it is currently the subject of debate or consensus building and by whom. In this, D-ADD provides organizations and employees with greater visibility into the agreement status of a constantly evolving D-Case.

9.3 OVERVIEW OF IMPLEMENTATION

Here we provide a brief overview of the implementation of D-ADD, which was implemented using the Java [12] and Scala [13] programming languages. More specifically, Java was used to realize the Fundamental Tools and the Hybrid Database, while the Core was programmed mainly using Scala.

Figure 9-8 shows the class for the D-Case Model. D-Case is stored in the graph database via the Perpetuation Section. Similarly, the Consensus Building Support Tool has been implemented to provide the consensus building methods described in Subsection 9.2.2. Using this tool, nodes are added to corresponding models as consensus is built. Furthermore, using the decomposition patterns described in Section 4.6, D-Cases for different types of target systems are systematically handled. For example, as the Architecture Decomposition pattern handles D-Case for each subsystem, all of the models related to the D-Case can be stored together in D-ADD for efficient storage utility and execution.

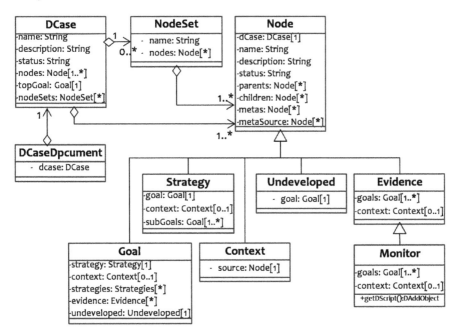

Fig. 9-8 Class for D-Case Model.

Rules in the Rule Processing part have been implemented as objects of the classes shown in Fig. 9-9. These rules are executed whenever a model state changes (i.e., the rule is satisfied) and the corresponding process is run. Rules can be written so as to include references to model information from the Model Definition section. In addition, plugins for extending model-status acquisition can be used in rules—a typical example being a plugin for statistical analysis.

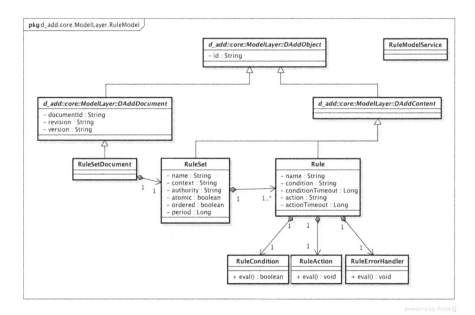

Fig. 9-9 Class for Rules in the Rule Processing Part.

In regard to the Model Definition section, we have defined, for example, a Consensus-building model to support the process of consensus building, a D-Case model for recording D-Case description, a Target model for monitoring targets, a Rule model, and a Deployment model for dynamically deploying rules and scripts in D-ADD.

In the Hybrid Database, various data associated with D-Cases that play a crucial role in achieving accountability are recorded using a graph framework. The implementation in question uses the open-source TinkerPop [14] solutions for the graph framework, Neo4j [15] (also open source) as the graph database, and the web application framework Play! [16] to achieve RESTful access. Play! does not support connection to a graph database, and therefore, we have configured TinkerPop to provide functionality for accessing Neo4j from Play!. Play! modules provide a means for adding plugins, and this functionality has been utilized for model and graph database connections. Meanwhile, the Fundamental Tools have also been implemented as Play! applications.

9.4 USING D-ADD IN PRACTICAL BUSINESS SITUATIONS

The range of potential applications for D-ADD is extremely wide, and in this section, we consider system development using the DEOS Process and D-ADD. It is very common for Java application servers to be used in large-scale information systems, and in web systems in particular. This approach is adopted in order to support distributed development at all stages from design to operation as well as code management, updates, and management of the overall system configuration; furthermore, it also makes a range of maintenance functions available for use during system operation. The dependability of the system under development can, therefore, be adversely affected to a significant degree if application servers are not selected correctly based on its characteristic features.

Nowadays, communication between stakeholders is often recorded during the course of large-scale IT development processes making use of application servers. The ad-hoc approaches of the past often depended on mailing lists for information sharing, but recent years have seen a shift towards Redmine and other tools that provide project-management functions in order to systematically manage and utilize communications, records, evidence, and reports. The minutes of meetings with users and other stakeholders are also recorded by these tools, and this data is now frequently referred to when major specifications changes must be implemented. The DEOS Process could be seen as a set of techniques for dealing with quality and dependability at certain times in the future, but in this specific field, there are no other solutions for linking stakeholders' agreements from the development phase through to the operation phase of the information system throughout its lifecycle. As such, the DEOS Process represents the world's first experiment of its kind.

Figure 9-10 organizes D-ADD and the related business according to a forward model and a reverse model. The forward model represents application to a new development project, and it allows the entire DEOS Process to be

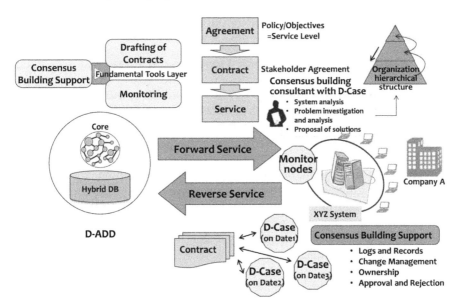

Fig. 9-10 D-ADD and Related Business.

integrated into the information system in question. In this case, D-ADD functionality could conceivably be utilized for business based on the SaaS model, business providing D-Case support for consensus building, and applications business built around the Fundamental Tools.

In accordance with TOGAF® [17], meanwhile, architecture repositories that support the Enterprise Continuum are established and used to store the deliverables of the Architecture Development Method process. D-ADD is an ideal solution for implementing these architecture repositories.

The reverse model, on the other hand, represents application of the DEOS Process to an existing system. Here, D-ADD could be used for services that support this type of application, services for diagnosing system weak points in terms of OSD assurance, and D-Case development services for existing systems.

ITIL [18] is a set of best practices for IT service management. Under this approach, strategies for transforming raw data successively into *Information*, into *Knowledge*, and ultimately into *Wisdom* are organized based on the ITIL Service Lifecycle framework. Incorporating the ITIL approach into the D-ADD Model Processing part allows its best practices to be utilized in DEOS Process implementation.

Java remains the core technology for application servers, but producers of other commercial packages and open-source software are working hard to make inroads. However, none of the solutions currently available are well suited to the OSD assurance techniques proposed by the DEOS Project, and this may ultimately provide many opportunities to apply D-ADD in practical business situations. As information systems play an evermore important role in the infrastructure of society, proper implementation of the DEOS Process using D-ADD can pave the way for significantly enhancing OSD.

9.4.1 Thoughts on Business Domains for DEOS

As will be described below, we have identified three business domains that could benefit greatly from application of the DEOS Process—namely, education, medicine, and design engineering. Common to all three are an expansive knowledge system and a firm basis in academic information. Debate in each of these disciplines can often dissolve into trivialities, leading to the corresponding arguments becoming distorted or disjointed. Information must be reconfigured in response, but steps taken to fill the gaps and to also connect and combine information can cause further distortion. This ubiquitous modification, updating, and reconfiguration of knowledge is often a latent characteristic of information systems, and the DEOS Process for assurance of dependability is particularly applicable in such a case.

(a) Education

Let's take a look at one possible application of the DEOS Process in the field of education. In this field, the curriculum is a highly important matter that concerns educators, managers, students, and their guardians and affects all teaching activities. Recent government[2] policies encouraging full disclosure of syllabuses aiming at disseminating university knowledge are just some of the efforts being made to make curricula more transparent. Certain organizations seek to strictly manage the syllabus as a contract between the educator and the guardian, and any modification thereof is treated as a significant change in the curriculum.

While it is critical that the description of a curriculum or syllabus is defined rigidly in terms of meaning, these descriptions may contain ambiguities or inconsistencies or may not be structured in a comprehensive fashion. In light of this, we considered how the DEOS Process could be applied to practical curriculum or syllabus development. All concerned parties must reach consensus whenever students are drawing up a study plan and selecting a curriculum in line with its targets, and the DEOS Process could be introduced here. A study plan is an important element of a student's overall life plan, and it serves as a road map for four years of university study. Various interactions exist between curricula, and in a narrow sense, syllabuses must be interlinked and reference one another. Guardians bear the cost of educating the students they sponsor, and each will want to approve curriculum or syllabus choices from the perspective of the student's overall life plan. The university, meanwhile, must assure the quality of its education.

In this, the age of social networking, an education system premised on discussion and consensus building by educators and students could be implemented in line with the DEOS Process as common infrastructure shared and promoted by universities and other similar organizations all over the country, thereby providing a valuable opportunity to research next-generation learning media.

(b) Medicine

In the field of medicine, electronic health records have long been seen as standard IT solutions for physicians interacting with patients. Without doubt, they serve as a valuable tool for digital-information input and output by specialists making the best use of their medical expertise. These devices are typically connected to clinical databases or treatment-related data repositories containing huge amounts of diagnosis and dosage information, but in recent years, there has also been an increase in the number of services offering instant digitization and recording of patients' medical parameters using

---

[2]The Ministry of Education, Culture, Sports, Science and Technology, Japan.

wearable computers and so forth. According to people active in this field, epidemiological analysis of large volumes of information as epitomized by big data has been progressing at a rapid pace of late, and there is great potential for breakthroughs that far surpass results achieved to date.

That having been said, very few information systems have actually been configured specifically for on-site diagnostic procedures or based on the medical expertise of physicians. This stems from the fact that medical information systems originated in hospital administrative procedures and have not developed beyond this level. Furthermore, electronic health records have not been fully implemented because—from the perspective of the physician— they are not necessarily of benefit in every situation.

In order to effectively develop massive, knowledge-centric information systems involving a great many physicians and others with valuable personal expertise, it will be necessary to realize services that provide for more rational consensus building by further enhancing D-ADD, storing not only the system's design information but also important knowledge content from other stakeholders, and thereby making the solution more comprehensive overall. Specifications defining what physicians require of clinical databases in terms of functionality and interlinking with other related medical data would then be well aligned with systematized medical knowledge, and any breakthrough made possible by such an approach would likely contribute greatly to advances in medicine.

(c) Design engineering

The DEOS Process could be applied to extremely long-term, specialized design projects involving, for example, airplanes, huge buildings, or massive production plants. In this type of extended project, consideration must be given to the potential for changes affecting the design or specifications, replacement of designers, budgetary adjustments, revisions to applicable laws, and environmental changes. As the DEOS Process accommodates changes in requirements with respect to the initial plan, it is ideal for this type of situation. Outside the realm of information systems, meanwhile, it can also be used effectively in the execution of manual tasks. In fact, a certain major telecommunications company introduced to the DEOS Process has expressed an interest in using this approach to control and manage organization work involving several thousand employees.

When constructing IT systems to support large-scale operations, a consulting company will often take the first step of analyzing the operations in question, identifying any problems, and proposing a To-Be model so that a high quality system can be realized. However, when IT vendors start to work with the consulting company's design, implementation problems and inconsistencies in the details may be identified, meaning that large-scale design changes must be made. This type of subcontracting to multiple parties in the development of information systems can have a particularly damaging effect on system dependability. The DEOS Process and D-ADD are clearly well suited to application as standardization tools for the development of large, enterprise-type information systems.

9.4.2 D-ADD as an Enterprise Repository

D-ADD could also conceivably be used as part of a corporate information system. In such a case, the target system would be the corporation itself, meaning that corporate activities aimed at securing profit would constitute sub-processes.

In order to provide support for consensus building as part of corporate decision making, D-ADD could store all directly related information and as much ancillary data as possible. In other words, it would be required to maintain information closely associated with valuable corporate resources in the form of employees, property, and finances. The relationships with two specific sets of information—namely, organization and authorization—are of particular importance in this regard. The organization encompasses personnel, but these two entities do not share the same authorization. These direct concepts and information activities in the organization space are liable to lead to inconsistency, disagreement, discrepancies, omissions, or in the worst-case scenario, illegal processes.

After financial window-dressing on a massive scale at Enron became a scandal of global proportions, the United States responded by enacting the Sarbanes-Oxley Act (SOX). In Japan, meanwhile, the Financial Instruments and Exchange Act (often referred to as "JSOX") required listed companies to enforce internal controls when conducting business. Penalties for violation of these laws even include delisting, and this has led to strict controls on approval processes, authorization checking, and user management, particularly within business systems. Internal operation rules are also controlled to prevent illegal processes, and furthermore, companies are now required to have mechanisms for verifying that no such processes are being executed.

In order for D-ADD to be put to use throughout the entire corporation, corporate resources that evolve over time must be appropriately managed; furthermore, the accumulation of data by the combined D-ADD and resource management must support the corporate structure and accommodate changes therein. Together, they must be defined as an enterprise repository, and the structure of the corporation should be designed as *architecture* for this reason as TOGAF®, for example, stipulates.

9.5 D-ADD & SOFTWARE DEVELOPMENT PROCESS INNOVATION

In this section, we look at how the conventional development process for large-scale IT system can be transformed by the application of D-ADD. In the development of these systems, an assurance case in the form of D-Case notation, D-ADD support for consensus building, and the accountability achievement functionality provided by D-ADD can enhance purposive quality much more than the design and production quality achievable with conventional methods. Particularly in the case of financial, medical, and other large-scale IT systems that need to be highly reliable, a waterfall-type development process has generally been considered to be preferable in order that a large number of stakeholders can come together to debate performance-related and other requirements over an extended period of time.

However, if operational requirements are extremely complex and the client also needs the system to be adaptable in the future, an iterative development process is often tried. Agile development that aims to identify the required specifications is sometimes employed at the start of such an iterative process. One of the problems with the waterfall approach is that a huge volume of documentation is produced over a long period of time and with the involvement of many parties; consequently, the system attributes detailed in these documents may not always reflect a unique set of specifications. This can make it difficult to identify inconsistencies in the specifications.

With agile development, however, the fact that iterative development is a shorter process means that a limited number of people will be involved, and therefore, it is highly likely that the specifications will be well defined. While we must assume that specifications will always have gaps and omissions, the iterative development process can efficiently address these issues, and as such, constitutes a methodology for enhancing completeness. Ultimately, the development process itself will be called into question in both of these cases in regard to how accurately communication is being recorded.

It has been some time since the updating of cellular-phone firmware has become possible via networks using the OTA (Over-The-Air) update technology. Furthermore, the recent proliferation of cloud services has provided users with customized services and also made possible the alteration thereof, all thanks to bandit algorithms [19]. In such a development environment, a new method called DevOps (development and operations) [20] has become the focus of considerable attention.

DevOps is an approach to system building where close communication is maintained between development and operations teams. With this type of system, the real substance of specifications often becomes apparent not while developing, but during operation. This type of situation is already covered by the DEOS Process, as D-ADD allows all stakeholders to confirm consistency between the agreements and the system's operation in real-time. Existing software development comprises requirement, development, operation, and other processes, arranged from upstream to downstream, and this format is well suited to software as a closed system. However, the modern development styles mentioned above need to envisage software as an open system. The DEOS Process and D-ADD provide full support for these styles to become standard in the development and operation of various applications and social infrastructures.

9.6 CONCLUSION

Within the overall DEOS project, this database was developed last of all so that it would play an important role in integrating the other technologies being developed. Several prototypes were created, each using the preceding prototype versions. We have integrated all of the required D-ADD functionality into the latest version, and it is available in the form of a web application from the DEOS website.[3]

To verify online performance as a database for D-Case storage, we connected D-ADD to the D-Case editing tool AssureNote, which was described in Section 8.4. We are gaining valuable experience and know how through ongoing use by multiple simultaneous users and involving multiple D-Cases.

REFERENCES

[1] http://www.opengroup.org/subjectareas/enterprise/archimate
[2] The Japan Commercial Broadcasters Association, Broadcasting Handbook, Revised Edition, Nikkei Business Publications Inc., Tokyo, 2007 (in Japanese).
[3] NEC Corporation, Broadcasting Solutions: Commercial Broadcast System, http://jpn.nec.com/media/hoso/cmwin.html (in Japanese).
[4] Unixon Systems, Co., Ltd., Introduction to digital.HOX, http://www.unixon.co.jp/product_dhox.html (in Japanese).
[5] Ministry of Internal Affairs and Communications, 2013 Telecommunications White Paper, 2013 (in Japanese).
    http://www.soumu.go.jp/johotsusintokei/whitepaper/ja/h25/pdf/index.html
[6] Inohara, T. 2011. Consensus Building, KeisoShobo, Tokyo (in Japanese).
[7] Nishida, K. 1990. An Inquiry into the God, New Haven, CT: Yale University Press.
[8] Shimizu, H. 1995. Ba-Principle: New Logic for the Real-time Emergence of Information, Holonics, 5/1 (1995): 67–69.
[9] Nonaka, I. and N. Konno. 1998. The concept of 'Ba': Building foundation for Knowledge Creation, California Management Review, Vol. 40, No. 3, Spring.
[10] Kinoshita, Y. and M. Takeyama. 2013. Assurance Case as a Proof in a Theory: towards Formulation of Rebuttals, in Assuring the Safety of Systems—Proceedings of the Twenty-first Safety-critical Systems Symposium, Bristol, UK, 5–7th February 2013, C. Dale and T. Anderson (eds.). SCSC, pp. 205–230, Feb. 2013.

[3]http://www.jst.go.jp/crest/crest-os/osddeos/index-e.html

[11] Kinoshita, Y., M. Takeyama, M. Hirai, Y. Yuasa and H. Kido. 2013. Assurance Case Description using D-Case in Agda, D-Case in Verification and Validation, National Institute of Advanced Industrial Science and Technology, Japan, pp. 1–18.

[12] http://www.java.com/ja/

[13] http://www.scala-lang.org/

[14] http://tinkerpop.com

[15] http://neo4j.org

[16] http://www.playframework.org

[17] The Open Group, TOGAF® Version 9.1, ISBN: 978-9-0875-3679-4, 2011.

[18] The IT Service Management Forum, ITIL® Foundation Handbook, ISBN: 978-0-1133-1349-5, The Stationary Office, 2012. http://www.tsoshop.co.uk

[19] White, J.M. 2012. Bandit Algorithms for Website Optimization, O'Reilly Media, December 2012.

[20] Loukides, M. 2012. What is DevOps? Infrastructure as Code, O'Reilly Velocity Web Performance and Operations Conference, O'Reilly, June 2012.

10

STANDARDIZATION OF OPEN SYSTEMS DEPENDABILITY

10.1 STANDARDIZATION AS A BASIC TECHNOLOGY FOR OSD

Modern information processing systems are invariably interconnected either directly or indirectly to other systems, which constitute the *environment*. This environment could be the Internet, or alternatively, it could be a set of components used by the system in question. Because such a system is crucially affected by the dependability of the other systems in the environment, it makes no sense to focus only on the dependability of the system under consideration. A hands-off approach would not work here.

That having been said, the environment cannot be clearly defined and usually changes over time. In that sense, information processing systems are typical of open systems. As soon as a definite boundary is imposed upon it, something develops outside that boundary and demands attention. This is an instance of the OSD *uncertainty* that will be discussed in Section 10.3.

In order to achieve dependability despite such uncertainty, the dependability of both the system under consideration and the environment must be assured in a specific objective fashion through the adoption of standardization. Take the issue of security as an example: assuming that perfect security is unattainable and there is a limit on how much can be spent to achieve it, we must live with a certain—but not the highest—level of security. Who, however, decides what is a good compromise in terms of security level? Particularly when faced with an indefinite set of stakeholders, the system developer simply cannot estimate the degree of security associated with most, if not all, of them.

There are no objective theories that tell us how to estimate the degree of security. The adequacy of any given measure is determined by factors such as the stakeholders' values and social norms, meaning that estimation goes beyond simple scientific facts concerning the system's mechanisms. Standards can play a role here. For instance, security standards, which are defined on the basis of universally accepted theories and social common sense, can determine the degree of security that would be acceptable to stakeholders in a given set of circumstances.

The environment of the system under consideration is often defined by socially accepted concepts, and as those concepts usually have many interpretations, it tends to be ambiguous in nature. Yet we need a clear definition of the environment in order to achieve dependability of the system. Applying standardization and reorganization as clearly defined rules in order to clarify the vague and implicit evaluation criteria shared by society would serve as an important starting point in the pursuit of Open Systems Dependability. Clear definitions achieved through standardization would also make it much easier for stakeholders from different backgrounds to communicate effectively with one another in terms of both the system and its environment.

We cannot separate the dependability of the system under consideration from that of its environment. All stakeholders, therefore, must communicate with one another in order that dependability may be achieved. This is not a moral lesson, but a conclusion drawn from scientific and logical observation. Standardization represents a core technology in the quest for OSD. In this context, it is not a means of spreading established technological advances; rather, standardization is one of the steps that must be taken in order to achieve Open Systems Dependability.

10.2 STANDARDIZATION: SOCIAL AGREEMENT ON MINIMUM MEASURES

In this section, we overview some examples of difficulties experienced in separating the dependability of a system from that of the environment. We also consider the role of standardization in overcoming such difficulties.

10.2.1 Shared Estimation of Dependability Integrity Levels and Process Standards

The following are typical concerns of system stakeholders:

- "What dependability integrity level is required of the system in order to be accepted by society? We don't want to incur excess cost by going beyond what is necessary with dependability management relative to other similar systems."

- "In order to devise our own dependability management plan, we need to be informed of the dependability management plans of related systems in the environment. It would not make sense for a significantly higher level of dependability to be achieved in our system when compared with that achieved in the related systems."

- "We, the development team, realize that our dependability management plan should be integrated with that of the operation team; however, their conventions are very different from ours, and that makes integration difficult. Moreover, we may be forced to follow their approach as our team is significantly smaller. That would also be a problem."

We can see that communication would clearly benefit if development and operations teams, organizations, and society as a whole were to share common conventions for dependability management. Standards could contribute to such sharing of conventions. In order to achieve this, however, concepts concerning processes and products as well as terms and vocabulary need to be standardized.

The Information-technology Promotion Agency (IPA) developed and maintains the Common Framework for system life cycle processes [15]. This framework preceded ISO/IEC 12207 [12] and ISO/IEC 15288 [13] which adopts many of the ideas developed therein.

10.2.2 The Need for Standardized Communication

Regardless of the extent of deliberate consideration, not all system problems can be prevented.[1] This underlines the fact that dependability communication between system stakeholders is crucial.

Inadequate dependability communication can cause difficulties between the supplier (including developers, operators, and maintainers) and the acquirer. Complaints such as the following are not uncommon for suppliers.

(1) "I really can't believe that the system we provided would be used like that! It's not our fault if it breaks down as a result."

(2) "System B has incredibly detailed conditions in terms of usage, and our System A does not comply. We won't accept the blame for ignoring such absurd requirements!"

(3) "I simply cannot believe that System B depended on this function of our System A. It is not guaranteed in the system specifications, and changes were made to the functionality with the new version of System A in order to improve performance. The results were disastrous."

---

[1]Needless to say, this is no excuse for inadequate risk treatment.

Systems cannot be expected to operate correctly under all circumstances. Every system comes with preconditions for normal operation, and failure to satisfy them often causes problems. It is, however, impossible in practice to write down all of these preconditions precisely. In Case (1) above, the acquirer uses the system in a way the supplier did not foresee. In Case (2), the supplier does what the acquirer did in Case (1). Here, the acquirer makes another contract in which he plays the role of the supplier. The system covered by the second contract provides a service to be used by the system covered by the first, and these two systems are interconnected with one another. The problem in Case (3) does not arise in the development process, but instead in the maintenance process. A potential consequence of miscommunication in the initial development revealed itself upon system renewal.

We must understand that foreseeing and enumerating all such possible failures in advance—not to mention taking countermeasures against them—is fundamentally impossible. Whenever potential damage is known, preventive steps can be considered; furthermore, action can be taken to avoid recurrence if damage is actually done. It is simply not possible, however, to implement countermeasures for *all* conceivable types of damage, whether expected or unexpected.

The issue here is how to provide for damage that is not foreseen in advance. Because unforeseen damage can never be explicit until revealed, post-failure treatment is essential. Maintaining close communication among stakeholders—particularly, between the supplier and the acquirer—would work effectively to improve this.

The following are typical problems on the acquirer side:

- An acquirer experiences failure of his system and says, "All of our subcontractors insist that they are not responsible because others *may* be in charge. They all lack any responsibility whatsoever!"

- A subcontractor might remark, "I am not sure of the extent to which we must give an account of this failure to satisfy the customer. However, I am concerned that a lot of confidential information may be disclosed in such an explanation."

- A client in the process of developing a dependability management plan might say, "We simply wish to know what would happen in the case of specific failures, but the supplier refuses to give straight answers. We are starting to lose confidence in them, but who else can we trust?"

The acquirer needs to resolve the problems caused by the system failure, so he or she seeks

1. assistance from the supplier for the treatment, and moreover,
2. compensation for the damage (depending on the extent of the supplier's responsibility for the failure).

Both of the above are, needless to say, contrary to the interests of the supplier. Close communication in advance of any failure in order to define the extent to which the damage should be treated would help the parties get out of this difficult situation.

As a tool for this sort of communication, the assurance case is a documented body that details measures for implementing dependability with respect to all presumed risk, as well as the argument for why those measures are appropriate. The dependability communication process can be made explicit and evaluable by defining the format for the assurance case and the way it should be written. Two standards for doing so are described below—namely, ISO/IEC 15026-2 [6] and OMG SACM [20].

10.2.3 Communication Quality & MetaAssurance Case Standardization

In the previous section, we discussed the importance of communication among stakeholders. Mere communication, however, is not sufficient—the ideas must be communicated *correctly* and *precisely* in order to achieve Open Systems Dependability. A stakeholder could comment on shared ideas as follows:

- "This off-the-shelf Component A came with an assurance case that meets our own criteria. However, the acquirer of our System B, which utilizes Component A, did not accept our assurance case for System B because the assurance case for Component A did not meet their criteria. Apparently, the acquirer's criteria for assurance cases differ from ours. This is a problem because rewriting the assurance cases for all of the components used in our system for each acquirer would be far too costly."

In some cases, a partner may not be very serious about dependability communication:

- "We wish to reach a consensus with our partner in regard to the treatment of failure. However, they respond with throwaway, idealistic comments without engaging in any concrete discussion. For example, they offer to do 'their best' to avoid failure or say that stop gap measures will suffice because failure cannot occur as long as their basic assumptions hold. I expect that a state of chaos will develop in the event of any failure, particularly an unexpected one. We may have no option but to seek a legal judgment in order to resolve this impasse."

We would need to share this concept throughout the entire system, not only with fixed partners, but also with potential ones. One way to do so would be to write down explicitly the basic ideas concerning dependability as a standard.

Making a fuss upon failure is never a smart move. Providing is preventing, and it is wise to prepare in advance. However, establishing a common understanding of the fundamentals of systems and dependability with each and every partner (including potential ones) is not practical. It would be difficult to achieve OSD through this approach.

Evaluating the system's assurance case is one effective way of determining whether dependability communication is progressing smoothly. The appropriateness of the assurance-case evaluation itself may be argued, and yet another assurance case may be written. The latter assurance case is known as the metaassurance case.

Research and standardization in the field of assurance-case evaluation have recently been progressing more or less in parallel. Standardization of the requirements for metaassurance cases will help us to achieve a common vision regarding how assurance cases should be evaluated.

10.3 INHERENT UNCERTAINTY OF OPEN SYSTEMS AND STANDARDIZATION

By their very nature, open systems contain uncertainty. As we shall see in this section, standardization of the interfaces between systems could help to resolve the problems that this causes.

As noted in Section 10.1, standardization is one of the basic technologies for achieving OSD. It plays an active role, as opposed to a passive one whose purpose is merely to establish the compatibility of system tools and components. In the latter case, standardization of a technology begins after it has been developed and completed to a certain degree. Here, technical development and standardization are activities that can be separated from one another. This approach is based on the premise that technical development can be carried out in a laboratory (*in vitro*) and may then be continued in line with actual conditions, adjusting the technology where necessary. *In vitro* development should be seen as a closed-system activity.

This closed development approach alone cannot resolve the open system problems discussed in the previous section. The presence of unforeseeable blackbox elements, which are shrouded in secrecy, forces the development process to go outside the closed laboratory. Development processes in open-system conditions need to interact with the environment, and standardization is one effective means of realizing this exchange. It does so by distributing standards throughout society, with the corresponding changes in society giving existence to the concepts defined therein.

Let us look at the example of commercial off-the-shelf (COTS) components. Suppliers of these products rarely disclose technical details in order to protect their intellectual property. In such a system, the COTS portion is a blackbox, and thus, the system integrator utilizing COTS components cannot fully control the dependability of the system. To make matters worse, providers can freely upgrade COTS components whenever they see fit as long as the user interface is kept as is, and this further restricts what the integrator can control.

One might think that COTS manufacturers should be compelled to disclose their technologies in detail. That would, after all, give the system integrator complete control over the system's dependability. This approach, however, does not work in practice. Firstly, it would be extremely difficult to convince manufacturers to take the risky step of disclosing their intellectual property by providing details of their products. Secondly, such disclosure would require enormous cost. Thirdly, but most importantly, this approach would ultimately negate the need to use COTS solutions in the first place. Thus, any attempt to force COTS manufacturers to disclose the technical details of their products would be fruitless.

This type of blackbox is just one example of the uncertainty, incompleteness, and indeterminacy that are characteristic of open systems. These characteristics are also found in network services, the reuse of legacy code, and changes in the system environment and market. All of these encompass uncertainty in an inability to foresee how and where change will occur, incompleteness in being unable to confirm all details, and other issues such as the lack of a single actor managing all related systems.

Instead of being forced to provide all technical details, the suppliers themselves could proactively provide evidence for assurance that their COTS *achieve a certain level of dependability.*

Trying to achieve dependability of a system in isolation does not work. This is because the system of interest is ultimately connected to many other systems in the environment, and its dependability also hinges on that of the other systems. In terms of systems ecology, this may be regarded as a fact.

There is also an example of security in daily life. Even if we follow the rules and mind our own business, our safety is at risk whenever people who break the law are nearby. If everyone were to behave, however, the security and safety of all could be guaranteed. Similarly, dependability in a system containing blackbox sections cannot be achieved through the efforts of the blackbox users alone. It must be achieved in all related systems, including the blackbox sections.

In order that evidence for the assurance of dependability may be provided, a fixed dependability framework for the provision thereof must be shared. To reach the goal of dependability not only in one's own system but also in all related systems, we must define what dependability actually entails and the conditions for achieving it. These conditions are nothing less than dependability requirements. The aim of IEC 62853, which is introduced below in Section 10.4.1, is to establish an OSD requirement standard.

Dependability requirement standards serve as an important tool for achieving dependability in open systems, including systems using COTS components, network services, and legacy code.

10.4 STANDARDIZATION ACTIVITIES CONCERNING DEOS

In this section, we provide an overview of standardization activities relating to DEOS technologies.

The technologies developed as part of the DEOS Project can be divided into two groups. The first—including the DEOS Process, D-Case notation, and other related elements—sets forth requirements for achieving OSD by redefining system lifecycle processes. The second group comprises the techniques needed to realize these processes—notably D-ADD, D-RE, D-Script, D-Case Editor, and D-Case in Agda. The DEOS standardization strategy calls for standards to be established for each of these groups (Fig. 10-1).

Requirement standards related to OSD take the form of *de jure* standards developed and maintained by the ISO/IEC JTC1/SC7 *Software and systems engineering* committee and the IEC TC56 *Dependability* committee.

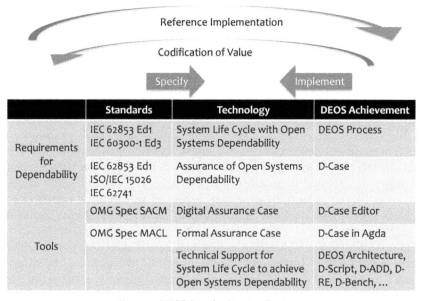

	Standards	Technology	DEOS Achievement
Requirements for Dependability	IEC 62853 Ed1 IEC 60300-1 Ed3	System Life Cycle with Open Systems Dependability	DEOS Process
	IEC 62853 Ed1 ISO/IEC 15026 IEC 62741	Assurance of Open Systems Dependability	D-Case
Tools	OMG Spec SACM	Digital Assurance Case	D-Case Editor
	OMG Spec MACL	Formal Assurance Case	D-Case in Agda
		Technical Support for System Life Cycle to achieve Open Systems Dependability	DEOS Architecture, D-Script, D-ADD, D-RE, D-Bench, ...

Fig. 10-1 DEOS Standardization Strategy.

The mission of JTC1/SC7 is to establish and maintain standards for the engineering of software projects and systems, but in recent years, it has become increasingly concerned with standards for systems and software assurance, which have attributes very relevant to dependability. Meanwhile, TC56 has been developing and maintaining international standards for reliability and dependability for over four decades, and while its community has extensive experience in the fields of plant and mechanical engineering, they do not include many participants from software engineering.

Tool standards related to OSD are being developed by *de facto* or *forum* standard bodies, such as the System Assurance (SysA) Task Force at the Object Management Group (OMG) and The Open Group (TOG).

10.4.1 Requirements Standards

This section provides an overview of standardization work concerning requirements for achieving OSD. The aim of these activities is to establish international standards for the requirements a system must satisfy in order to achieve OSD.

In Section 10.2, we looked at three kinds of standards—namely, process standards, assurance case standards, and meta-assurance case standards— and these are currently being developed (Fig. 10-2). Process standards have already been formulated or are currently being established in the form of ISO/IEC 15026-4:2012 [8], which deals with assurance requirements, and IEC 62853 [3], which is under development as of February, 2015. ISO/IEC 15026-4 was established by adding assurance process views to ISO/IEC 12207 [12] and ISO/IEC 15288 [13], which define software life cycle processes and system lifecycle processes, respectively. Assurance case standards have either already been established or are now under development in the form of ISO/IEC 15026-2:2011 [6] and IEC 62741 [2]. In addition, efforts are under way to realize metaassurance case standards, also in IEC 62853.

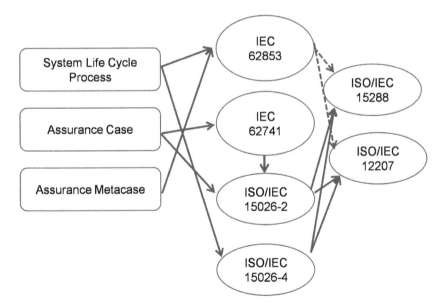

Fig. 10-2 Three Aspects of DEOS Technology and International Standards.

(1) System assurance requirements standards

System assurance and dependability are closely related terms. ISO/IEC 15026 *Systems & Software Assurance*, which comprises four parts dealing with system and software assurance, was published in full in 2012. This standard was developed by, and is now maintained by, ISO/IEC Joint Technical Committee 1 (JTC1): *Information Technology*, and more specifically, by its Working Group 7 (WG7): *Life Cycle Management* at the committee's Subcommittee 7 (SC7): *Software and Systems Engineering*.

Although this standard was originally published in 1998 under the title *System Integrity Level*, it was renamed *Systems and Software Assurance* upon review and comprises four separate parts. The original standard has been preserved as Part 3; furthermore, Part 4, which was published in September 2012, completes this standard. ISO/IEC 15026, as it is now known, can be seen as a totally new standard, independent of the earlier 2008 version. Part 1, *Concepts and Vocabulary* [5], was first released in 2010 as a technical report; however, this was reformatted as an international standard and it defines the concepts and vocabulary for the other parts.

Part 2, titled *Assurance Case* [6], defines the format and content of the assurance case. The *safety cases* referred to in the *Automobile Functional-Safety Standard* ISO 26262 and the *Security Target Documents* (also called security cases) that play a key role in the *Common Criteria* (ISO/IEC 15408 [9, 10, 11]) are instances of assurance cases. ISO/IEC 15026-2 specifies what assurance cases are.

Part 3 of ISO/IEC 15026 was published in 2011 under the title, *System Integrity Levels* [7]. Here, the *System Integrity Level* is generalized for system states such as safety and security from the *Safety Integrity Level* (*SIL*) used in the functional-safety standard IEC 61508 and the *Evaluation Assurance Level* (*EAL*) used in the *Common Criteria* (ISO/IEC 15408 [9, 10, 11]). This is similar to generalization of the assurance case from the safety case and the security case.

Not limited to system attributes such as safety and security, ISO/IEC 15026 Part 3 (and its original 1998 version, which was translated as JISX0134 [14]) also establishes methods for defining the integrity levels of the system under consideration with respect to other system attributes such as reliability, availability, and integrity.

Part 4, *Assurance in the Life Cycle*, serves as a guideline that establishes matters required for achievement of dependability in each process of the overall lifecycle of a system or software.

ISO/IEC JTC1 develops and maintains standards concerning the lifecycle of systems and software, such as ISO/IEC 15288 *System life cycle processes* and ISO/IEC 12207 *Software life cycle processes*. Rather than focusing only on lifecycle processes in the narrow sense of architecture design, implementation, operation, maintenance, and decommissioning as elicited from requirements, each of these standards also encompasses a wide range of other related processes such as procurement, service provision, and project management.

ISO/IEC 15026-4 recommends activities that are crucial to achieving assurance in terms of the more important aspects of the lifecycle processes under each of the two standards mentioned above.

(2) OSD requirement standards

The requirements that must be satisfied in order to achieve OSD for accommodating change and uncertainty were studied by the DEOS Project. Based on this, a new work item proposal to develop an International Standard of requirements for OSD was submitted by the Japan's national committee of IEC TC56 in 2012. The proposal was approved and IEC TC56 Project Team 4.8 (PT4.8) was formed accordingly, whose mission is to develop IEC 62853/Ed.1 [3]. As of October 2014, comments for the first committee draft (1CD) thereof were collected from the national committees, and the second committee draft (2CD) is anticipated early 2015. The standard is expected to be published in December 2016.

10.4.2 Tool Standardization

We now turn our attention to standardization of the tools required for OSD in the various system lifecycle processes. Standards for these tools include the specifications of the tools themselves and formats for input and output data.

(1) OMG SACM (Structured Assurance Case Metamodel)

The OMG Structured Assurance Case Metamodel (SACM) [20] defines a data format for assurance cases, and D-Cases developed using D-Case Editor conform to this standard. Preparing assurance case files so as to conform makes it easier to convert data for use with other tools. SACM was published in February 2013.

OMG (the Object Management Group) is a standardization consortium that provides UML and CORBA standards among others, and its activities are centered in the United States. The group's System Assurance Task Force (SysA TF) develops and maintains technical standards for assurance cases and other assurance-related technologies.

(2) OMG MACL (Machine-checkable Assurance Case Language)

SysA TF is currently involved in preparations for a Machine-checkable Assurance Case Language (MACL) standard—in specific terms, this concerns assurance case notation languages that facilitate mechanical verification of assurance case integrity (i.e., automatic checking thereof by a computer).

The D-Case in Agda tool prototyped as part of the DEOS Project for the structuring and checking of D-Cases is not limited to syntactical verification but also extends to semantic checking. D-Case in Agda has been implemented in Agda, a functional programming language with dependent types based on constructive type theory. Any other language designed on the basis of this theory would also suffice.

The development of MACL is still at a very early stage, but it will probably provide an abstract syntax for machine checkable assurance cases, leaving concrete syntax undetermined so that many of the existing languages for assurance case description could easily be incorporated. OMG issued a Request for Information (RFI) concerning MACL in September 2012, and work is currently under way on the subsequent Request for Proposals (RFP) that will define the details of the specification proposals to be submitted.

10.4.3 TOG Dependability through Assuredness™ Framework

At a general meeting held in July 2013, The Open Group announced publication of its *Dependability through Assuredness™ Standard* [18] (O-DA) for real-time and embedded systems, and this is based on the DEOS Process concept. This marked the fruition of our work with the TOG Real-time Embedded Forum over the course of more than two years.

The normative elements of the O-DA standard are set forth in Section 3: *The Dependability through Assuredness™ (O-DA) Framework*, and the framework is defined as comprising the following five elements (Fig. 10-3):

(1) Dependability Modeling

(2) Assurance Case Development

(3) Accountability

(4) Failure Response Cycle

(5) Change Accommodation Cycle

In the Dependability Modeling phase, the architect must appreciate the need for the system to be dependable, and various risk dependencies are modeled in order that the assurance case can be developed. The O-DA standard proposes using TOG's *Dependency Modeling (O-DM)* standard [23] for this purpose. In the next phase, Assurance Case Development, the actual assurance case is developed in order that the requirements of the system under consideration may be assured. The O-DA standard presents a set of recommended steps for achieving this.

As in the DEOS Process, accountability is a main element of the O-DA standard. To achieve accountability, the timing of responsive actions and the parties who are to execute them must be defined based on the assurance case developed in the preceding phase, and the stakeholders are required to agree

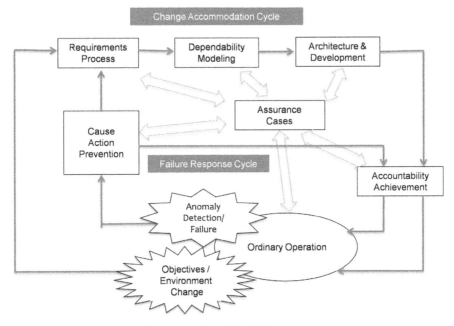

Fig. 10-3 Outline of the O-DA Standard.

on these definitions. This takes place in the Accountability phase, which forms a part of both the *Failure Response Cycle* and the *Change Accommodation Cycle*.

As in the DEOS Process, the *Failure Response Cycle* is initiated in order to execute actions concerning deviation of the information system from operation standards agreed upon by the stakeholders. The purpose of these actions can be to prevent deviation from occurring or to recover when it does; furthermore, the cause of any deviation is analyzed in this cycle. The *Change Accommodation Cycle*, meanwhile, is initiated in response to changes in the system's environment, and its aim is to achieve accountability through requirements elicitation, dependability modeling, development of the system's architecture, and system development.

As a guideline for implementing the O-DA framework, the informative part of the standard describes links with other standards (FAIR [24], SACM [25], and SBVR [26]), relations to existing techniques for assurance case notation, the development of a high-confidence assurance case, and the use of formal methods. An appendix includes case studies of application of the O-DA standard to the TOGAF® [22] ADM and to the DEOS Process and architecture. TOGAF® presents guidelines for developing an architecture that can accommodate change at the Enterprise Architecture (EA) level. In contrast, the O-DA standard addresses change accommodation at the system architecture level, but the two standards complement one another to support application of the DEOS Process in practical situations.

10.5 CONCLUSION

Standards related to OSD are currently being developed from the twin perspectives of requirements standardization and tools standardization. The ISO/IEC JTC1/SC7 *Software and Systems Engineering* and IEC TC56 *Dependability* committees are working on *de jure* requirements standards, while OMG, TOG, and others are developing forum standards and technical specifications for tools standardization.

REFERENCES

[1] Tokoro, M. (ed.). 2012. Open Systems Dependability—Dependability Engineering for Ever-Changing Systems, CRC Press.
[2] IEC 62741/Ed.1 Guide to the demonstration of dependability requirements. The dependability case, International Standard, 2015-2-17.
[3] IEC 62853/Ed.1 Open Systems Dependability, Work in progress.
[4] ISO/IEC 15026:1998 IS Information Technology—Software Engineering—System Integrity Level (superseded by [7]).
[5] ISO/IEC 15026-1:2013 IS Information Technology—Software Engineering—Systems and software assurance—Part 1 Concept and vocabulary.
[6] ISO/IEC 15026-2:2011 IS Information Technology—Software Engineering—Systems and software assurance—Part 2 Assurance case.
[7] ISO/IEC 15026-3:2011 IS Information Technology—Software Engineering—Systems and software assurance—Part 3 System Integrity Level.
[8] ISO/IEC 15026-4:2012 IS Information Technology—Software Engineering—Systems and software assurance—Part 4 Assurance in the life cycle.
[9] ISO/IEC15408-1:2009 Information technology—Security techniques—Evaluation criteria for IT security Part 1 Introduction and general model.
[10] ISO/IEC15408 Information technology—Security techniques—Evaluation criteria for IT security Part 2 Security functional requirements.
[11] ISO/IEC15408 Information technology—Security techniques—Evaluation criteria for IT security Part 3 Security assurance requirements.
[12] ISO/IEC 12207:2008 IS Information Technology—Software Engineering—Software life cycle processes.
[13] ISO/IEC 15288:2008 IS Information Technology—Software Engineering—System life cycle processes.
[14] JIS X0134-1999: Information technology—System and software integrity levels, Japanese Industrial Standards Committee (in Japanese).
[15] Common Frame 2013—Towards realization of Usable Systems together with managers and business departments, Software Engineering Center (SEC), Engineering Division, Information-technology Promotion Agency (IPA), 2013 (in Japanese).
[16] Kinoshita, Y. and M. Takeyama. 2013. OSD International Standardization Strategies for Practical Application of DEOS, DEOS-FY2013-IS-01J (in Japanese).
[17] Embedded Technology 2012 Special Session C8, Pacifico Yokohama, http://www.jst.go.jp/crest/crest-os/osddeos/event/201211/et2012.html, November 16, 2012 (in Japanese).
[18] The Open Group Standard, Real-Time and Embedded Systems: Dependability through Assuredness™ (O-DA) Framework, The Open Group, 2013.
[19] The Open Group Standard, Risk Taxonomy, The Open Group, 2009.
[20] Object Management Group Standard, Structured Assurance Case Metamodel (SACM), Version 1.0, OMG Document Number: formal/2013-02-01. http://www.omg.org/spec/SACM

[21] Object Management Group Standard, Semantics of Business Vocabulary and Business Rules (SBVR), v1.0, OMG Document Number: formal/2008-01-02. http://www.omg. org/spec/SBVR/1.0/PDF

[22] The Open Group Standard, TOGAF® Version 9.1, The Open Group, 2011.

[23] The Open Group Standard, Dependency Modeling (O-DM), Constructing a Data Model to Manage Risk and Build Trust between Inter-Dependent Enterprises, The Open Group, 2012.

[24] The Open Group Standard, Risk Taxonomy, The Open Group, 2009.

[25] Object Management Group Standard, Structured Assurance Case Metamodel (SACM), Version 1.0, OMG Document Number: formal/2013-02-01. http://www.omg.org/spec/SACM

[26] Object Management Group Standard, Semantics of Business Vocabulary and Business Rules (SBVR), v1.0, OMG Document Number: formal/2008-01-02. http://www.omg. org/spec/SBVR/1.0/PDF

11

CONCLUSIONS

11.1 RETROSPECT

Upon its establishment in 2006, the DEOS Project began research in the field of dependable operating systems for embedded systems aiming at practical applications. The project underwent several changes thereafter, before delivering the findings described in these pages.

The first of these changes involved the definition of an embedded system. At the beginning of the project, this term was commonly interpreted as meaning a single application running on a terminal with a small number of sensors and actuators; however, the embedded system we needed to address in reality was an application on a terminal operating in conjunction with a back-end server. As a result, we needed to extend the definition accordingly. Systems with this type of configuration are numerous and include mobile phones, rail and air ticket gates, electronic points-of-sale in supermarkets, and ATMs. Extending the scope of the project in this way meant that server systems and distributed systems would also need to be considered, so we decided that the term "embedded systems" should be replaced with "real-world systems". The *Dependable Embedded OS* developed as the first deliverable of the DEOS Project supported this extension by providing enhanced functionality for hierarchical isolation of processes and for monitoring and logging.

The second major change affecting the DEOS Project began with doubt as to whether dependability could be assured with an OS alone. Following this, there was much discussion as to whether any system could be confidently declared dependable and who would be responsible if a system claimed to be absolutely dependable actually failed. In today's world, the number of systems that continue to operate indefinitely with no changes to their specifications is far surpassed by the number with ongoing structural and functional modifications to accommodate change in the objectives of service providers or in the system environment. The DEOS Project members shared a recognition that dependability would also have to be assured in systems of this type. This led to the decision to focus our research on open systems that exist and operate in changing environments. Ultimately, we concluded that dependability in open systems was achievable only by means of an iterative process, and also that stakeholders' agreement and the achievement of accountability within such a process were essential for dependability. The concept of Open Systems Dependability was thus established.

To put this concept into practice, we defined the DEOS Process as an iterative process executed within a double loop comprising a *Failure Response Cycle* and a *Change Accommodation Cycle*. Considering that the system may need to be modified depending on the outcome of failure response, the double loop of the DEOS Process includes the highly distinctive feature of a path from the *Failure Response Cycle* to the *Change Accommodation Cycle*. This double-loop process ensured compatibility with the dependable embedded OS that was already under development and was renamed the DEOS Runtime Environment (D-RE).

In parallel with these efforts, various arguments were put forward regarding who would be a stakeholder and what shareholder agreement would actually entail. In addition, we introduced the concept of assuredness. Specifically, we developed an extension of the assurance case called the D-Case, making it suitable for use during system development and modification and also during system operation, and we adopted it as a means for consensus building in the DEOS Process. We supplemented D-Case with D-Script to provide scripting functionality for monitoring, recording, and control during system operation, thereby enabling flexible emergency response in the event of a failure or a sign of impending failure. Recognizing that agreement logs in D-Case and D-Script as well as operation logs of system monitoring and actions could be used effectively in achieving accountability, we defined and developed the Agreement Description Database (D-ADD) to make this possible. What emerged was the DEOS Architecture, defined as (1) the DEOS Runtime Environment (D-RE), which includes support for security, (2) the DEOS Agreement Description Database (D-ADD), which retains all the agreement logs and operation logs, (3) consensus-building tools, and (4) tools for the development of application software.

Thanks to this experience in dealing with major change, the DEOS Project was able to configure a new technological system (or paradigm) that recognizes accountability achieved on the basis of consensus building as the essence of dependability in all phases of the system lifecycle from development through operation. To distinguish this new technological system from prior dependability techniques, we named it *Dependability Engineering for Open Systems* (DEOS).

The DEOS Project sought not only to achieve groundbreaking results, but also to make those results applicable in practical situations. With this in mind, a Research Promotion Board with members from industry was established at the very beginning to drive the project, and this board kept us aware of real-world needs as research and development work proceeded. Our plan was to deliver the project's findings not simply in the form of scholarly papers but also as software that could be put to actual use, and we established the DEOS R&D Center (Appendix A.1) for this purpose. We further realized that international standardization and the establishment of a user consortium would be necessary if our findings were to be applied to actual systems all over the world. We began work aimed at international standardization from around the middle of the DEOS Project, and as a result of this effort, The Open Group (TOG) industrial standards organization published the *Dependability through Assuredness*™ *(O-DA) Framework*, a standard based on our research, in July 2013. In addition, the International Electrotechnical Commission is in the process of standardizing our findings as IEC 62853 *Open Systems Dependability*. A user consortium, the Association of Dependability Engineering for Open Systems (DEOS Association; Appendix A.2), was established in October 2013 after more than a year of intensive planning. We also held a series of Workshops on Open Systems Dependability (WOSD), and the fact that a fourth will be held in November 2014 indicates growing international recognition of this approach to dependability.

The pioneering results achieved by the DEOS Project are due primarily to the efforts of the research teams, but also to the support of the Research Promotion Board and many other individuals and organizations, and we are deeply grateful to everyone who has contributed. More must be done so that our achievements may be put to practical use. At present, the DEOS Runtime Environment (D-RE), a number of associated development tools, and D-Case tools are freely available for download; meanwhile, a version of the D-ADD agreement description database has been released as a web application. In terms of practical use, D-RE has, for example, been used in the development of a biped robot designed to interact with humans and in other systems; D-Case is being used in the development of systems for engine control, satellites, and electrical power exchange among others; and we look forward to further applications going forward. What's more, D-Case training sessions are held on a frequent basis, and the number of attendees is on the increase.

11.2 PROSPECTS

In the course of developing the DEOS Process and D-Case based on the Open Systems Dependability concept, the DEOS researchers have become increasingly confident that the DEOS technological system can be applicable to a wide range of systems. The reasons are twofold:

(1) Many systems are operating for extended periods of time in changing environments, yet their dependability is based on conventional methods. These systems need a new base for dependability in order that accountability may be achieved, and this is where the DEOS technological system can contribute.

(2) These systems are built with the types of complex software with which the DEOS technological system has proved essential and effective.

With these thoughts in mind, new application domains including intelligent transport systems, cloud servers, disaster management planning, and medical systems are being considered, and the DEOS technological system will be broadened and extended in response. The support of our readers, who come from wide ranging fields, is of great importance in achieving dependability in complex, ever-changing systems, and for this reason, in contributing to the establishment of a dependable society.

APPENDIX

A.1 THE DEOS PROJECT

A.1.1 Goals & Outline

The DEOS[1] Project was launched in October 2006 as one area of research within the CREST Strategic Basic Research Program at the Japan Science and Technology Agency (JST), seeking *dependable operating systems for embedded systems aiming at practical applications*. Given that the majority of today's embedded systems are connected to servers via networks and services are provided to users by the entire integrated system, we targeted our research not only toward operating systems for embedded systems in the narrow sense of the word, but also toward system software that would provide dependable operation, the tools needed to develop that software, and the corresponding processes for system development and operation.

Modern information systems are massive, complex, and must accommodate ongoing change in their objectives and environments. As such, it is more appropriate to perceive them as *open systems* than as closed. It was concluded early in the DEOS Project that an iterative approach must be adopted in order to achieve dependability in this type of system, and the basic concept of *Open Systems Dependability (OSD)* was born. The goal of the project could then be defined as follows:

> To develop a methodology and specific development means for dependable system software that will ensure the continuous delivery of services to users by appropriately adapting systems to ongoing changes in their objectives and environment. The fundamental concept for this methodology is *Open Systems Dependability (OSD)*, and the development means comprises the iterative *DEOS Process*, the *DEOS Architecture* that enables this process, and the technologies required to implement this architecture.

The *OSD* concept and the *DEOS Process* will be applicable in the development not only of dependable system software, but also a wide range of massive, complex systems that must accommodate change. In other words,

---

[1]DEOS was originally an abbreviation of Dependable Embedded Operating Systems.

this process can be put to use with open systems in general. Hence, the abbreviation DEOS stands for Dependability Engineering for Open Systems.

A.1.2 The R&D System for the DEOS Project

Research work under the DEOS Project was led by a research supervisor and a deputy research supervisor, who were supported by several area-specific advisors. Together, they advised the research teams as to the direction their R&D should take and evaluated their progress. Area management advisors provided advice on policies for the practical implementation of research findings and other matters. In addition, managers and engineers from industry interacted with the research teams as members of the research promotion board, leveraging their real-world experience to help guide the R&D findings toward practical implementation.

The project got under way in 2006 with five teams, and a further four were added in 2008. The project teams established in 2006 primarily studied elemental technologies such as virtual machines, server group virtualization, systems software verification, benchmarking, fault simulation, and real-time low power consumption systems. This involved investigation of the types of system to be considered in the project and refining the concept of dependability in this regard. The teams added in 2008 supported these activities, ultimately leading to development of the DEOS Process and the DEOS Architecture. They also started research in fields such as requirements analysis, processes for consensus building and notation, and security, and they joined in the preparation of international standards based on the project's findings. The 2006 teams were funded by JST until March 2012; the 2008 teams, until March 2014.

A core team was also set up in 2008 with members from the various research teams. In parallel with the work of the individual teams, it has overseen all DEOS Project research, refining the R&D approach in order to remain focused on the ultimate goal. In 2010, the members were split into sub-core teams to work in particular areas with the corresponding project teams. These sub-core teams oversaw productive R&D on major components of the *DEOS Process* and the *DEOS Architecture*, including D-Case, D-Script and monitoring, virtual machines and multiple operating systems, system software verification, and DS-Bench/Test-Env (Fig. A-1).

Research work by the original 2006 project teams came to an end in March 2012, after which the 2008 teams took the reins and initiated Process & Architecture Committee (Fig. A-2). Research was principally focused on the following six areas: D-Case, D-Script, D-ADD, Security, D-RE and DEOS application, and Verification and Standardization.

The DEOS R&D Center worked with industry to evaluate the findings of the project and research teams and apply them to actual products and services. In order to do so, they integrated the research achievements of the different teams, factored in consideration of intellectual property rights and

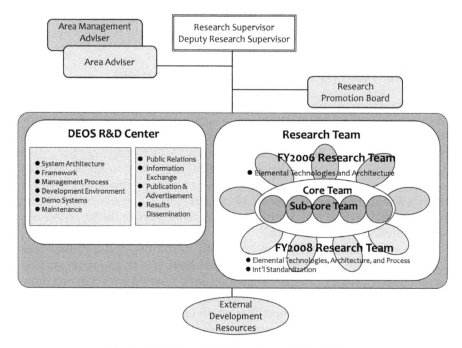

Fig. A-1 DEOS Project Configuration until March 2012.

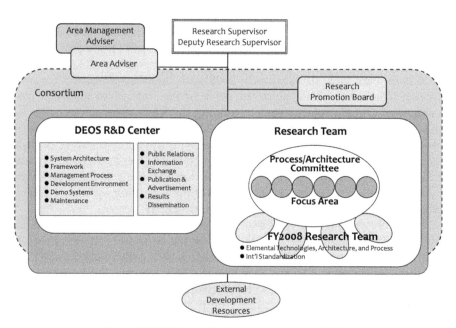

Fig. A-2 DEOS Project Configuration since April 2012.

maintainability, carried out testing, evaluated readiness for use, and packaged the resulting software. The software realized through our research can be freely downloaded from the DEOS Project web site (http://www.jst.go.jp/crest/crest-os/osddeos/index-e.html) (refer to Section A.1.5)

A.1.3 DEOS Project Roadmap

The following phases represent major milestones in the DEOS Project research (Fig. A-3):

- **Phase 1** (October 2006 to September 2009): Establishment of the dependability concept. Presentation of the system architecture with development and operation processes supporting the concept and including key evaluation indices. Demonstration of a reference system implementation that integrated the elemental technologies of the research teams launched in 2006.

- **Phase 2** (October 2009 to September 2011): Implementation of the system architecture together with D-RE and tools incorporating the elemental technologies developed by the 2006 teams. Initial canvassing of industry and research institutes for participation in an evaluation consortium. Preparation for international standardization of key findings. Demonstration of D-RE and tools incorporating the elemental technologies from the research teams launched in 2008.

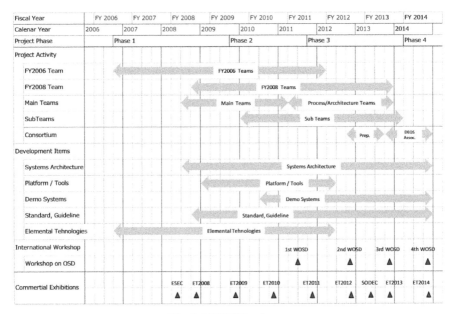

Fig. A-3 DEOS Roadmap.

- **Phase 3** (October 2011 to March 2014): Release of software and other deliverables. Trial implementation and assessment of the deliverables by industry and other users, and further development for practical use based on feedback. Establishment of the consortium. Preparation for industrial standardization of the DEOS concept, and international standardization based on the DEOS Project findings.
- **Phase 4** (April 2014 onward): Implementation, adjustment, and further development of the project's findings through the efforts of international standards organizations, industrial standards organizations, and the consortium.

A.1.4 Principal DEOS Project Members

Research Supervisor		
	Mario Tokoro	Executive Advisor & Founder, Sony Computer Science Laboratories, Inc.
Deputy Research Supervisor		
	Youichi Muraoka	Professor Emeritus, Waseda University
Area Advisors		
	Kazuo Iwano	Advisor, Business Service Group, Mitsubishi Corporation
	Toru Kikuno	Professor, Faculty of Informatics, Osaka Gakuin University
	Koichi Matsuda	Advisor, Information-technology Promotion Agency
	Koichiro Ochimizu	Vice-President, Japan Advanced Institute of Science and Technology Specially-Appointed Professor, Research Center for Highly Dependable Embedded Systems Technology
	Yoshiki Seno	Chief Technology Specialist, Knowledge Discovery Research Laboratories, NEC
	Hidehiko Tanaka	President, Dean & Professor, Institute of Information Security, Graduate School of Information Security
	Hiroto Yasuura	Trustee & Vice President, Kyushu University Professor, Graduate School of Information Science and Electrical Engineering, Kyushu University System LSI Research Center Manager, Kyushu University
2006 Team Research Leaders		Area
	Yutaka Ishikawa	Center Manager and Professor, Information Technology Center, The University of Tokyo	Dependable single system image operating system for parallel/distributed embedded systems
	Toshiyuki Maeda	Assistant Professor, Graduate School of Information Science and Technology, The University of Tokyo	Development technologies for dependable system software

Tatsuo Nakajima	Professor, Faculty of Science and Engineering, Waseda University	Dependable operating system for high-performance information appliances
Mitsuhisa Sato	Center Manager and Professor, Center for Computational Sciences, University of Tsukuba	Calculation platform for dependable embedded parallel systems with low power consumption
Hideyuki Tokuda	Professor, Faculty of Environment and Information Studies, Keio University	Dependable operating system for micro ubiquitous nodes
2008 Team Research Leaders		Area
Satoshi Kagami	Assistant Manager, Digital Human Research Center, National Institute of Advanced Industrial Science and Technology	Real-time, parallel dependable OS and distributed networks thereof
Yoshiki Kinoshita	Professor, Department of Information Science, Kanagawa University	User-oriented dependability
Kenji Kono	Associate Professor, Faculty of Science and Technology, Keio University	Producing highly attack-resistant secure operating systems
Kimio Kuramitsu	Associate Professor, Faculty of Engineering, Yokohama National University	Assuring continuous security during execution using Security Weaver and P-Scripts
Research Promotion Board members		
Nobuhiro Asai	Distinguished Engineer, Tokyo Software Development Laboratory, IBM Japan	
Shingo Kamiya	Software Engineering Group Manager, Framework Business Unit, Solutions Division, NTT Data IntelliLink Corporation	
Tadashi Morita	Visiting Researcher, Sony Computer Science Laboratories, Inc.	
Masamichi Nakagawa	Global Solutions Development Group Manager, R&D Division, Panasonic Corporation	
Takeshi Ono	Systems Platform Group Manager, PA Systems Department, Systems Division, IA Platform Business Headquarters, Yokogawa Electric Corporation	
Ichiro Yamaura	No. 2 Controller Platform Development Group Manager, Controller Development Division, Fuji Xerox Co., Ltd.	
Area management advisors		
Kazuo Kajimoto	Systems Engineering Center Manager, Panasonic Corporation	
Yuzuru Tanaka	Professor, Graduate School of Information Science and Technology, Hokkaido University	
Tetsuya Toi	Executive Officer, Fuji Xerox Co., Ltd.	
Seishiro Tsuruho	Principal, HAL Tokyo	
DEOS R&D Center		
Makoto Yashiro	DEOS R&D Center Manager, Japan Science and Technology Agency	

(Names arranged in alphabetical order within groups; affiliations of 2006 Team research leaders are as of March 31, 2012; others as of October 31, 2013)

A.1.5 DEOS Project Reports, Links & Software

Project plans and reports (DEOS Project documentation; in Japanese)

DEOS-FY2009-WP-01J:	DEOS Project White Paper Version 1.0
DEOS-FY2010-WP-02J:	DEOS Project White Paper Version 2.0
DEOS-FY2011-WP-03J:	DEOS Project White Paper Version 3.0
DEOS-FY2012-PU-01J:	DEOS Project Update 2012
DEOS-FY2013-SS-01J:	DEOS Project Research Achievements

Focus area technical reports (DEOS Project documentation; in Japanese)

DEOS-FY2013-DC-02J:	D-Case—Methods & Tools for Dependability Consensus Building
DEOS-FY2012-DS-01J:	D-Script—Failure Response with Scripting
DEOS-FY2013-DA-02J:	Agreement Description Database—Repository for Linking Open Systems Dependability and D-Case
DEOS-FY2012-SD-01J:	Base System Software for Extending Service Lifespan
DEOS-FY2012-RA-01J:	D-Case Robot Application Case Study—Mobile Miraikan Robot
DEOS-FY2012-SV-01J:	Writing Assurance Cases with D-Case in Agda Tool
DEOS-FY2013-IS-01J:	OSD International Standardization Strategy for Practical DEOS Implementation

Explanatory material produced for project system and software (DEOS Project documentation; in Japanese)

DEOS-FY2013-PR-01J:	DEOS Programming Reference (with overview of reference implementation)
DEOS-FY2013-RE-01J:	D-RE Specifications with D-RE Implementation Guide
DEOS-FY2013-CW-01J:	D-Case Weaver Specifications with Implementation and Usage Guide
DEOS-FY2013-BT-01J:	DS-Bench/Test-Env Execution Guide
DEOS-FY2013-EA-01J:	D-RE API Specifications
DEOS-FY2013-EC-01J:	D-RE Command Specifications
DEOS-FY2013-SP-01J:	D-RE API Sample Program Guide
DEOS-FY2013-BI-01J:	DS-Bench/Test-Env Specifications
DEOS-FY2013-BI-01J:	DS-Bench/Test-Env Environment Build Procedure

DEOS-FY2013-BC-01J:	DS-Bench and D-Case Editor Interface Specifications
DEOS-FY2013-VS-02J:	D-Visor86 and D-System Monitor Environment Build Procedure
DEOS-FY2013-QK-01J:	QEMU-KVM and D-System Monitor Environment Build Procedure
DEOS-FY2013-DI-01J:	Demo System Install Manual
DEOS-FY2013-MD-01J:	D-Case—Collaboration with Modeling Environment; Demonstration Materials
DEOS-FY2013-MP-01J:	D-Case—Collaboration with Modeling Environment; Plug-in Installation Manual
DEOS-FY2013-MI-01J:	D-Case—Collaboration with Modeling Environment; Environment Build Procedure
DEOS-FY2013-MT-01J:	D-Case—Collaboration with Modeling Environment; Tutorial

Published books

ISBN: 978-1-46657-751-0	*Open Systems Dependability* (CRC Press)
ISBN: 978-4-86293-079-8	*An Introduction to D-Case* (Daitec Holding; in Japanese)
ISBN: 978-4-86293-091-0	*Let's Use Practice D-Case Dependability Cases* (Asset Management Co., Ltd.; in Japanese)
ISBN: 978-4-7649-0461-3	*DEOS: Dependability Engineering for Open Systems*, Kindaikagakusha, in Japanese

DEOS Project web site:	http://www.jst.go.jp/crest/crest-os/osddeos/index-e.html

Software downloadable from the DEOS Project website:
D-Case Editor
D-Case Weaver
D-Case Stencil
D-Case/Agda
D-RE 1.0
DEOS Reference System
D-Visor & D-System Monitor
System Recorder
D-Box
D-ADD (as a web application)
D-Case—collaboration with modeling environment
DE-Bench/Test-Env
Model Checking
SIAC (Single IP Address Cluster)

A.2 THE DEOS ASSOCIATION

To date, the DEOS Project has successfully developed a wide range of concepts, methods, systems, and tools with the aim of improving dependability, not only in embedded systems, but also in other information systems that must operate either continually for extended periods of time while accommodating change or are interlinked with systems operated and managed by others. In order to ensure that the achievements of our research and development efforts are widely applied and further enhanced, and that they contribute to enhanced dependability in information systems all over the world, the DEOS Center, the project's research teams, and the Research Promotion Board have provided support for establishment of the consortium and communicated with cooperating companies and organizations towards this end.

This preparation led to the founding of The Association of Dependability Engineering for Open Systems (the DEOS Association) on October 22, 2013. The association plans expansion of the DEOS approach into industry using patents, software, teaching material, books, and other research achievements of this project, and it promotes further research, development, verification, evaluation, and standardization of dependability technology with the aim of contributing to the safety, security, and comfort of society at large. The specific objectives are as follows:

- Encourage industry to apply the achievements of the DEOS Project;
- Promote further development of this work in line with the needs of industry and society;
- Help to enhance dependability in systems developed and operated by industry;
- Cultivate human resources capable of applying dependability technologies; and
- Support the creation of safer, more secure, and more comfortable information and communication infrastructures for society at large.

Figure A-4 shows the makeup of the DEOS Association at the time of its establishment. The DEOS Association can be found at http://DEOS.or.jp.

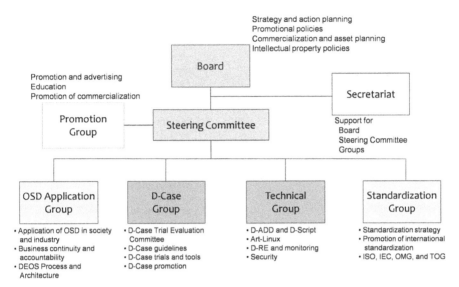

Fig. A-4 Makeup of DEOS Association.

A.3 EXAMPLES OF RECENT INFORMATION-SYSTEM FAILURES

	Date	Incident	Cause
1	January 2, 2013	Between 0:17 and 2:10, the nationwide 4G LTE data network in Japan could not easily be accessed, and up to 1.75 million calls were affected.	A signal control device incorrectly output a device alarm. Although the operator should normally have switched the device's card system in such a case, he initiated a complete device recovery process, causing all sessions with LTE handsets to be dropped. As a result, all of the handsets simultaneously tried to reconnect and extreme congestion occurred. The cause of the incorrect device alarm was a problem with software—it was configured such that normal conditions would be mistakenly identified as an anomaly. Repairs took until January 8, 2013.
2	December 31, 2012	Between 0:00 and 2:55, the 4G LTE data network operated by *au* could not be accessed, and from then until 4:23, users had difficulty accessing the network. Up to 1.8 million calls were affected throughout Japan.	The area of the network in which the data communication failure occurred comprised a base-station control device, a signal relay device, and a subscriber profile server for controlling traffic (up to 7 GB). In order to address any concentration of access attempts, the subscriber

	Date	Incident	Cause
			profile server had been provided with functionality for discarding signals from the various devices. At the time of the failure on December 31 (midnight on New Year's Eve), the number of access attempts was seven times the normal level. This triggered the server's countermeasure function, and responses to the relay device were consequently either non-existent or delayed and sessions were ultimately dropped. This resulted in the relay device resending the data or users trying to reconnect, exacerbating the congestion and rendering the LTE handsets unable to connect.
3	August 13 to 15, 2012	NTT DoCoMo mobile phones owned by 80,000 customers could not be used for voice calls or data communication in up to 220 national and regional domains. Roaming services did not function correctly.	Earlier that year in May, it was discovered that roaming services were only able to handle half of the rated traffic due to a mistake made when configuring a replacement device in March; however, it was thought that any repairs during the London Olympics could result in a large-scale failure, so this work was put off until later.
4	August 7, 2012	Network equipment from the Tokyo Stock Exchange derivatives trading system was struck by a hardware failure. Automatic switchover to backup equipment was attempted, but this was not possible due to the nature of the failure, and trading was interrupted.	Normally, the No. 2 backup device should have automatically replaced the main No. 1 device upon hardware failure, but a localized hardware failure was not correctly detected inside the No. 1 device, causing both the No. 1 and No. 2 devices to try to operate as the main device. As a result, the devices connected to the switch were unable to determine to which of the two devices to send data.
5	August 2 and 3, 2012	1.52 million users of NTT DoCoMo mobile phones had difficulty making calls.	One of two critical devices failed, and repairs had been put off because it was assumed that signals could be rerouted.
6	June 20, 2012	(1) The server configuration data and database content of a number of First Server customers was lost. (2) It is likely that confidential information was released onto the Internet during repair work.	(1) An updating program intended to address a vulnerability failed to operate correctly because code required to cancel a file deletion command was incomplete; in addition, a preventative function did not perform adequately due to checking in a test environment, and the incorrect files were deleted. (2) Problems with repair procedures and programs made it impossible to prevent the mixing of data belonging to different customers.

	Date	Incident	Cause
7	February 2, 2012	A data switching function of the Tokyo Stock Exchange securities trading system failed, preventing data from being delivered externally. As a result, trading of certain stocks was suspended for three and a half hours.	The system had triple redundancy, but when the first server failed, it mistakenly determined that switchover to the two backup servers had been successful and reported that failure response was complete. Because the switchover had actually ended in failure, the response was delayed and management was not notified.
8	Between June 6, 2011 and January 25, 2012 (five times)	Users of NTT DoCoMo mobile phones repeatedly experienced difficulties making calls or accessing data. Other problems such as email addresses being replaced with those of other customers also occurred.	The failures are thought to have been caused by malfunction of the system for managing handset position data, congestion of verification servers upon device switchover due to router malfunction, operational instability caused by incorrect traffic estimates made during the design of data converters that had just come into service, and other problems that occurred during this period.
9	April 21, 2011	Amazon EC2 and other services went offline. This failure also brought down Engine Yard, Heroku, and many other sites.	This problem was caused by incorrect network settings for Amazon EBS, an external storage service for virtual machines.
10	March 15 to 22, 2011	(1) Mizuho Bank overnight batch processing and online services were interrupted. (2) ATMs could not be used, and other problems such as currency exchange delays and duplication of bank transfers were also experienced.	The volume limit on transfers to an account for accepting disaster relief donations had not been set high enough in advance and was exceeded. This resulted in abnormal termination of overnight batch processing and a large number of associated operator errors.
11	August 10 to 12, 2010	Users were unable to access mixi—Japan's most-popular social networking site.	There was a *memcached* bug in a general-purpose distributed memory cache system. *The memcached* daemon suddenly stopped working when it had to deal with too many connections and disconnections.
12	May 22, 2009	JavaScript included in NTT DoCoMo mobile phones made it possible to access any site without authorization. The company had to suspend mobile phone sales.	Unauthorized access was possible because of a flaw in the JavaScript implementation. It is likely that there was a problem with the security implementation of the same-origin policy used by web browsers, and NTT DoCoMo is suspected of not having documented this fully in its mobile phone specifications.

	Date	Incident	Cause
13	February 24, 2009	Users of Gmail with Google Apps were unable to access their accounts.	An unforeseen service disruption occurred during routine maintenance at a data center. In this type of situation, users would be redirected to an alternate data center made available for maintenance work; however, new software for optimizing the location of user data had an unexpected side effect and triggered a latent bug in the Gmail code. As a result of this bug, the data center automatically switched to failure response when traffic was redirected to it, and this in turn caused multiple downstream overloads and overloaded the data center.
14	September 14, 2008	Check-in terminals at several airports stopped working, causing the cancellation of flights.	A certificate that authorized the terminals to access the server system had expired early that morning.
15	July 22, 2008	Data from a derivatives trading system could not be delivered to users of the system.	The working memory for each trade was set much smaller than the size deemed necessary, leading to the loss of several trades.
16	October 12, 2007	Automatic railway fare gates accepting IC cards stopped working in the Tokyo metropolitan area.	There was a rudimentary error in logic for splitting large blocks of data into smaller pieces upon the transmission of vital information from the server to the fare gates. This produced an infinite data-receive loop in the gates.
17	May 27, 2007	The All Nippon Airways check-in system stopped working, resulting in 130 flights being canceled and 306 delayed.	A problem with a network device caused by a hardware fault resulted in congestion between the host computers and terminals. The relationship between network device problems and congestion was not recognized and thus overlooked.
18	March 1, 2003	A system for processing flight-plan data went offline, causing the cancellation of 215 flights and the delay of 1,500 others.	This problem was caused by a single bug triggered when a certain area of memory was accessed; however, testing was insufficient for this bug to be detected in advance.

A.4 FACTORS IN OPEN SYSTEMS FAILURES

Classification

Incompleteness
System not completely constructed: It is often the case that the specifications initially required become inadequate, implementation does not fully match the specifications, and the system's behavior becomes difficult to ascertain fully at the time of shipment or service startup.

Underlying Factors

- Incomplete understanding of the system due to:
  - Increase in system size
  - Increase in system complexity
  - System connected to other systems
  - Use of open-source software
  - Use of black-box software
  - Use of legacy code
- Changes in system components
- Configuration changes

Specific factors

- ◆ A system combining numerous software components becomes too massive and complex to allow comprehensive specification description and testing
- ◆ An error or omission in specifications, design, implementation, or testing caused by discrepancies in the characterization of the system during the requirements management phase or development phase or by an error in documentation
- ◆ An error in updating or amending procedures for administration, operation, or maintenance or an error caused by the expiration of a license
- ◆ Inconsistency between the specifications and actual operation of open-source or black-box software components or legacy code, or poor understanding on the part of the developer
- ◆ Lack of certainty in the execution priority, sequences, and timing of modules
- ◆ Installation of components that were not planned for in the development phase or were not included in the test configuration, such as components downloaded from a network during operation.

Typical Symptoms

- Behavior not included in requirement documents
- Behavior not included in functional specifications
- Behavior not included in test specifications
- Unexpected behavior after the integration of components, even though each component passed functional tests
- Unexpected behavior after updating of a module during operation
- System stoppage without warning

Classification

Uncertainty
Change in the outside world with which the system is in contact: Changes to user requirements or the usage environment throughout the lifecycle of the system make it difficult to predict completely its behavior in the development or operation phases.

Underlying Factors

- Changes in expectations or human capabilities
- Changes in the operation environment
- Configuration changes
- Issues balancing system performance
  - Performance design
  - Capacity planning
- System becomes a network node

Specific Factors

- ◆ Changes in user requirements or expectations of the system, or changes in operator skill levels or capability during the maintenance and operation phases
- ◆ New modes of use due to increasing numbers of shipments or users, greater traffic, and changes in capacity economics
- ◆ Increased complexity due to onsite modification (manually or via the network) of component functions, changes of service, or reconfiguration of the system
- ◆ Changes in system resources (aging, memory restriction, CPU clock limitation, and so forth)
- ◆ Changes at the other end of the network (response or specifications changes)
- ◆ Unexpected connections or increasing interaction via the network, or malicious attacks from the outside

Typical Symptoms

- Processing speeds drop
- Longer waiting lines at user service terminals
- Wait messages displayed for services that are normally available instantly
- Acceptance of electronic money is suspended after the launch of a new service
- Failed attempts to transfer contributions or donations to bank accounts
- Introduction of 24-hour service is announced, but the system does not work during late-night hours
- Credit card numbers are stolen and used unlawfully
- System stoppage without warning

A.5 RELATED STANDARDS & ORGANIZATIONS

Standards

- IEC 61508: Functional safety of electrical/electronic/programmable electronic safety-related systems
  http://www.iec.ch/zone/fsafety/fsafety_entry.htm
- IEC 60300-1: Dependability management
  http://www.iec.ch/cgi-bin/procgi.pl/www/iecwww.p?wwwlang=E&wwwprog=sea22.p&search=text&searchfor=IEC+60300-1&submit=OK
- IEC 60300-2: Dependability Program Elements and Tasks
  http://www.iec.ch/cgi-bin/procgi.pl/www/iecwww.p?wwwlang=E&wwwprog=sea22.p&search=text&searchfor=IEC+60300-2&submit=OK
- ISO/IEC 12207: Systems and software engineering—Software life cycle processes http://www.iso.org/iso/iso_catalogue/catalogue_tc/catalogue_detail.htm?csnumber=21208

- ISO/IEC 15026: Systems and software engineering—Systems and software assurance http://www.iso.org/iso/home/store/catalogue_tc/catalogue_detail.htm?csnumber=62526
- ISO/IEC 15288: Systems and software engineering—System life cycle processes http://www.iso.org/iso/iso_catalogue/catalogue_tc/catalogue_detail.htm?csnumber=43564
- ISO 26262: Road vehicles—Functional safety http://www.iso.org/iso/catalogue_detail.htm?csnumber=43464
- IEC 61713: Software dependability through the software life-cycle processes—Application guide
  http://www.iec.ch/cgi-bin/procgi.pl/www/iecwww.p?wwwlang=E&wwwprog=sea22.p&search=text&searchfor=IEC+61713&submit=OK
- IEC 62347: Guidance on system dependability specifications
  http://www.iec.ch/cgi-bin/procgi.pl/www/iecwww.p?wwwlang=E&wwwprog=sea22.p&search=text&searchfor=IEC+62347&submit=OK

Process guides

- CMMI: Capability Maturity Model® Integration
  http://www.sei.cmu.edu/cmmi/
- DO-178B: Software Considerations in Airborne Systems and Equipment Certification http://www.rtca.org/
- MISRA-C:
  http://www.misra-c.com/
- TOGAF: The Open Group Architecture Framework
  http://www.opengroup.org/togaf/

Software

- SELinux: Security-Enhanced Linux
  http://www.nsa.gov/research/selinux/index.shtml
- AppArmor®: a Linux application security framework http://www.novell.com/linux/security/apparmor//
- Xen® hypervisor: the powerful open source industry standard for virtualization http://www.xen.org/

Related organizations and projects

- ISO: International Organization for Standardization
  http://www.iso.org/iso/home.htm

- IEC: International Electrotechnical Commission
  http://www.iec.ch/
- ISO/IEC JTC1: Joint ISO/IEC Technical Committee 1 http://www.iso.
  org/iso/standards_development/technical_committees/list_of_iso_
  technical_committees/iso_technical_committee.htm?commid=45020
- IEC/TC56: Technical Committee 56: IEC Technical Committee for
  International Standards in the field of Dependability
  http://tc56.iec.ch/index-tc56.html
- OpenTC Consortium: Open Trusted Computing Consortium
  http://www.opentc.net/
- Linux-HA Project: High Availability Linux Project
  http://linux-ha.org/
- Carrier Grade Linux Workgroup http://www.linuxfoundation.org/en/
  Carrier_Grade_Linux
- TCG: Trusted Computing Group
  http://www.trustedcomputinggroup.org/home
- ERTOS Group: Embedded Real-Time Operating-Systems Group http://
  ertos.nicta.com.au/
- ARTEMIS: Advanced Research & Technology for EMbedded Intelligence
  and Systems http://www.artemis.eu/
- CPS Program: Cyber-Physical Systems Program http://www.nsf.gov/
  pubs/2008/nsf08611/nsf08611.htm
- MISRA: Motor Industry Software Reliability Association
  http://www.misra.org.uk/
- AUTOSAR: AUTomotive Open System ARchitecture
  http://www.autosar.org/
- JasPar: Japan Automotive Software Platform and Architecture
  http://www.jaspar.jp/
- FlexRay Consortium: Consortium for the communications system for
  advanced automotive control applications
  http://www.flexray.com/
- The Open Group
  http://www3.opengroup.org/
- CoBIT: Control Objectives for Information and related Technology http://
  www.isaca.org/Knowledge-Center/COBIT/Pages/Overview.aspx
- ITIL: Information Technology Infrastructure Library http://www.itil.
  org/en/vomkennen/itil/index.php
- OMG: Object Management Group
  http://www.omg.org/

A.6 DEOS GLOSSARY

Accountability achievement: Dealing with system failure by explaining clearly to the stakeholders (specifically to users) the current status, the cause, and the recovery plan, or dealing with service changes and the like by explaining clearly to the stakeholders (specifically to users) the timing of restart and other conditions.

Action: A type of D-Case node that describes an operation procedure for responding to the corresponding failure.

Agreement Description Database: A database for storing the deliverables of the various stages of the *DEOS Process* for achieving agreements and accountability.

Anomaly detection: Detection of a state that may lead to a failure.

Assurance case: A document for assuring stakeholders of the validity of a claim.

Assure: Provide ample assurance that a claim is valid.

Assuredness: Ample assurance of the validity of a claim.

Availability: The ability of a system to maintain a high operating ratio.

Black box: A system or software component with unknown internal design and implementation that is integrated into a system only on the basis of external specifications.

Cause analysis: Identifying the cause of a failure based on agreements and evidence.

Change Accommodation Cycle: A cycle of the *DEOS Process* implemented in order to accommodate changes in stakeholder objectives or in the system environment.

Closed system: A system whose boundary, functions, and structure do not change.

Consensus building: A process whereby stakeholders sharing the same objectives reach mutual agreement by arguing specific positions.

Context: A type of D-Case node that provides information serving as assumptions for a goal or strategy.

DEOS Architecture: An architecture that supports the *DEOS Process* for a specific area or application and that comprises a DEOS Agreement Description Database (D-ADD), a DEOS Runtime Environment (D-RE), and various development tools.

DEOS Process: An iterative process for achieving *Open Systems Dependability* and that comprises the *Change Accommodation Cycle* and *Failure Response Cycle*.

Dependability: The ability to deliver continuously the services that are expected by users.

D-ADD: DEOS Agreement Description Database

D-Case: A method for building stakeholders agreement in the *DEOS Process* and the description of the resulting agreement itself.

D-DST: DEOS Development Support Tools

D-RE: DEOS Runtime Environment, which corresponds to what is commonly called an OS or kernel.

D-Script: A script-based DEOS technology that provides a flexible means of implementing failure management in an operational system.

Evidence: 1. Information that supports a particular claim in an argument. 2. A type of D-Case node that supports the goal in its ancestor node.

External: A type of D-Case node that represents a module managed by an external entity.

Failure: Deviation of an operational service from the service levels agreed by stakeholders.

Failure prevention: Prevention of system failure by detecting early signs thereof and other anomalies.

Failure Response Cycle: A cycle of the DEOS Process implemented in order to respond to system failure.

Formal verification: Proof of correctness or incorrectness of programs by formal or mathematical methods.

Goal: A type of D-Case node that describes a claim.

Incident: An event with adverse consequences.

Incompleteness: The inability to proof that a system can fully satisfy all requirements.

In-operation range: The range of parameters agreed by the stakeholders and within which an operational system is in the Ordinary Operation state.

Integrity: The ability of a system to prevent improper alteration of itself or its data.

Legacy software: Old software that still operates in an integrated fashion within a system and whose designers and maintainers are typically not available to perform maintenance.

Manage: Take action to resolve problems in order to keep the system in good condition.

Module: A type of D-Case node used to refer to the D-Case of another module.

Monitor: A type of D-Case node that provides evidence based on the system's operating state.

Ontology: A structure of concepts or entities within a domain, organized by relationships. More specifically, the vocabulary of a domain, the meanings of constituting terms, and their relations.

Open system: A system whose boundaries, functions, and structure change over time.

Open Systems Dependability (OSD): The ability of a system to accommodate change in its objectives and environment, to ensure that accountability in regard to the system is continually achieved, and to provide the expected services to users without interruption.

Open systems failure: A failure in an open system that is caused by incompleteness or uncertainty.

Ordinary Operation state: A state in which an operational system continually provides services at levels agreed by the stakeholders.

Parameter: A type of D-Case node that is used to set parameters within D-Case patterns.

Prevention of recurrent failure: Action or functionality to prevent a past failure or a similar one occurring again in the future.

Process: A sequence of steps or phases implemented in developing or operating a system or service.

Reliability: The ability of a system to deliver continually the expected performance over a specified period of time.

Responsibility: A type of D-Case node that explains the relationship between modules with different responsibility attributes.

Responsive action: Rapid, appropriate response to a system failure or an anomalous condition.

Requirements elicitation: Identification of requirements through the process of building consensus among stakeholders on objectives and needs.

Requirements management: A method for managing requirements based on stakeholders' agreement.

Serviceability (or *maintainability*): The ability to efficiently maintain a system through modification, debugging, and repair.

Security: Protection of a system from external attack that could impair availability, reliability, serviceability, and/or integrity.

Sign of failure: Manifestation of a system condition that could precede actual failure.

Specification description language: A language that describes the properties a program must have.

Stakeholder: An individual or organization who has a right, share, or interest in the service and/or system under consideration.

Stakeholders' agreement: Consensus among stakeholders concerning requirements and their implementation.

Strategy: A type of D-Case node that describes how a goal is to be argued in order to refine it by decomposing into sub-goals.

System architecture: A system's design concepts, fundamental functions, and basic structure.

Uncertainty: The inability to predict system changes and behavior in advance.

Undeveloped: A type of D-Case node that indicates that the argument or evidence is insufficient to support the goal.

LIST OF AUTHORS
(WITH CONTRIBUTING SECTIONS)

Atsushi ITO, Incubation Center, Fuji Xerox Co., Ltd.
Received M.S. from Graduate School of Media and Governance, Keio University. (§5.1)

Kiyoshi ONO, Researcher, Dependable Embedded OS R&D Center, Japan Science and Technology Agency.
Received Ph.D. from the University of Tokyo. (§7.1, §7.2)

Satoshi KAGAMI, Deputy Director, Digital Human Research Center, National Institute of Advanced Industrial Science and Technology.
Received Ph.D. from the University of Tokyo. (§7.3)

Yoshiki KINOSHITA, Professor of Information Science, Faculty of Science, Kanagawa University.
Received Ph.D. from the University of Tokyo. (§10)

Kimio KURAMITSU, Associate Professor, Division of Intelligent Systems Engineering, Faculty of Engineering, Yokohama National University.
Received Ph.D. from the University of Tokyo. (§8)

Kenji KONO, Associate Professor, Faculty of Science and Technology, Keio University.
Received Ph.D. from the University of Tokyo. (§7.4)

Hiroki TAKAMURA, Researcher, Dependable Embedded OS R&D Center, Japan Science and Technology Agency.
Received Ph.D. from Japan Advanced Institute of Science and Technology. (§2.1, §2.2, A.3)

Makoto TAKEYAMA, Researcher, Information Science, Faculty of Science, Kanagawa University.
Received Ph.D. in Computer Science from Edinburgh University. (§6, §10)

Hideyuki TANAKA, Researcher, Dependable Embedded OS R&D Center, Japan Science and Technology Agency.
Received M.S. from Oita University. (§5.2)

Mario TOKORO, Founder and Executive Advisor, Sony Computer Science Laboratories, Inc.
Received Ph.D. from Keio University. (§1, §2, §3, §11)

Tatsumi NAGAYAMA, Representative Director, Symphony Co., Ltd.
Received M.S. from Meiji University. (§9)

Yutaka MATSUNO, Assistant Professor, Graduate School of Information Systems, The University of Electro-Communication.
Received Ph.D. from the University of Tokyo. (§4.2, §4.3, §4.4, §5.3)

Shigeru MATSUBARA, Researcher, Dependable Embedded OS R&D Center, Japan Science and Technology Agency, Tokyo. (§2.2, §3.4, A.1, A.4)

Tomohiro MIYAHIRA, Researcher, Dependable Embedded OS R&D Center, Japan Science and Technology Agency.
Received M.S. from Tohoku University. (§7.1, §7.2)

Makoto YASHIRO, Director, Dependable Embedded OS R&D Center, Japan Science and Technology Agency.
Received M.S. from the University of Tokyo. (A.1, A.2)

Sachiko YANAGISAWA, Symphony Co., Ltd. (§9)

Hiroshi YAMADA, Associate Professor, Institute of Engineering, Tokyo University of Agriculture and Technology.
Received Ph.D. from Keio University. (§7.4)

Shuichiro YAMAMOTO, Professor, Information Communication Headquarters, Nagoya University.
Received Ph.D. from Nagoya University. (§4.1, §4.5, §4.6)

Yasuhiko YOKOTE, Representative Director and CDO, Cyber AI Entertainment Inc., and Project Professor of Graduate School of Media and Governance, Keio University.
Received Ph.D. in Computer Science from Keio University. (§7.1, §9)

INDEX